Quantitative Aptitude
Made Easy

Quantitative Aptitude
Made Easy

R.S.Priya
Director,
S.P Academic Consultancy
Chennai.

dīpti press
www.dıptipress.com

Copyright © 2015 www.diptipress.com

This book is exclusively marketed by
M/s Scitech Publications (India) Pvt. Ltd.,

This book or any part thereof may not be reproduced in any form without the written permission of the publisher.

978 81 8371 673 4

Published by V. Ramesh for dipti press, Chennai - 600 017

Preface

This book is developed keeping in mind the average and above average learners, who has the quest to achieve in the competitive exam.

This book meets the requirements of all the competitive exams conducted by various Boards.

Problems are specially coined to have a deep-in sight into the work of testing aptitude in mathematics.

All the problem are solved, and answers are arrived accurately. Hope, the learners would enjoy and appreciate the work. All the very best.

R.S.Priya

Contents

1. Numbers .... 1 - 12
2. HCF and LCM .... 13 - 22
3. Fractions: Decimal Fractions; Simplification Problems .... 23 - 32
4. Simplification .... 33 - 40
5. Powers and Roots .... 41 - 48
6. Word Problems on Numbers .... 49 - 57
7. Indices and Surds .... 58 - 63
8. Logarithms .... 64 - 67
9. Ratio and Proportion .... 68 - 73
10. Percentages .... 74 - 80
11. Averages .... 81 - 86
12. Ages .... 87 - 93
13. Profit and Loss .... 94 - 101
14. Business Mathematics - Partnership (Joint ventures) .... 102 - 109
15. Direct & Indirect Proportion .... 110 - 116
16. Time and Work .... 117 - 123
17. Pipes & Water Tanks .... 124 - 132
18. Time and Distance .... 133 - 139
19. Moving Trains .... 140 - 146

20.	Stream Crossing	147 - 152
21.	Geometry - Plane Figures	153 - 170
22.	Solid Figures	171 - 177
23.	Heights and Distances	178 - 181
24.	Algebra	182 - 189
25.	Alligation (Mixtures)	190 - 196
26.	Simple Interest (S.I)	197 - 202
27.	Compound Interest (C.I)	203 - 210
28.	Race and Skill Games	211 - 215
29.	Calendar	216 - 222
30.	Clocks	223 - 230
31.	Stocks and Shares	231 - 236
32.	True Discount	237 - 244
33.	Banker's Discount	245 - 250
34.	Non Conforming Items	251 - 254
35.	Permutations & Combinations	255 - 265
36.	Counting	266 - 274
37.	Probability	275 - 282

Data Interpretation

38.	Tabulation	283 - 291
39.	Bar Graphs	292 - 295
40.	Pie Charts	296 - 298
41.	Line Graphs	299 - 302

1

Numbers

Number Types

Real numbers: All the numbers on the number line are real numbers. All counting numbers $(1, 2, 3, \ldots 9)$, zero, & negatives of counting numbers are integers. 0 is neither +Ve nor -Ve. All numbers that can be expressed as the ratio of two integers are rational numbers.

All real numbers which are not rational (either +Ve or -Ve) are called irrational numbers: (eg, $\sqrt{2}, \pi, e$)

An integer that is divisible by another integer is a multiple of that integer: eg 14 is a multiple of 7.

The remainder is what is left over in a division problem. It is always smaller than the number we are dividing by:

For Ex.: 37 (dividend) $= 5 \times 7 + 2$ the number with which we are dividing is called Divisor; 7 is called the quotient; 2 is the remainder: $(2 < 5)$ **Factors** (or) the divisors of a number are the positive integers that evenly divide into that number. 48 has ten factors: $1, 2, 3, 4, 6, 8, \ldots 12, 16, 24$ & 48. They can be grouped in 5 pairs as follows. $[1 \times 48 = 2 \times 24 = 3 \times 16 = 4 \times 12 = 6 \times 8] = 48$

1	×	48
2	×	24
3	×	16
4	×	12
6	×	8

The **greatest common factor** (divisior)

GCF or HCF (Highest Common Factor) of a pair of numbers is the largest factor shared by the two numbers: For Ex.: the HCF of 36 & 48 is 12.

The Least Common Multiple of two numbers (LCM) is the least number which is exactly divisible by each one of the given numbers.

For ex. LCM of 36 & 48 is 144

(More of HCF & LCM in the following chapter)

Add in Div. Tests: Divisibility by 7: ex 343 subtract 2×3 from (34) we get 28: It is divisible by 7.

∴ 343 also is divisible number. If the resulting number is divisible by 7 then the original no is divisible by 7.

Divisibility Tests

-by 2 A number is divisible by 2. If its last (units) digit is divisible by 2.
 Ex: 158 is divisible by 2, also 170

-by 3 A number is divisible by 3, If the sum of the digits is divisible by 3
 For **Ex:** 47157 (Since sum = 24)

-by 4 A number is divisible by 4. If the last two digits are divisible by 4
 For **Ex:** 7184.

-by 5 If the units digit is 5 or 0
 Eg: 7185 or 7180

-by 6 If a number is divisible by 2 & 3 also, it is divisible by 6
 For eg 47454 (It is divisible by 2 since units digit is 4, (divisible by 2); sum of digits 24 divisible by 3 the number is divisible by 6)

-by 7 The digit in the units place is doubled & subtracted from the remaining number. If the resulting no is divisible by 7, the no given also will be div. by 7

-by 8 If the number formed by the last three digits is divisible by 8. 739168 is divisible by 8 since 168 is divisible by 8.

-by 9 If the sum of the digits is divisible by 9. For eg 739368 is divisible by 9 since the sum of the digits is 36, divisible by 9.

-by 10 If the units digit is 0.
 Ex: 798380

-by 11 A number is divisible by 11, if the difference of the sum of the digits in the odd places & that in the even places is either zero or divisible by 11.
 $9\ \overline{7}\ 6\ \overline{3}\ 2\ \overline{4}\ 5\ \overline{8}\ 1\ \overline{0}\ 8\ \overline{0}\ 2$ is divisible by 11 since.
 $9 + 6 + 2 + 5 + 1 + 8 + 2 = 33$ & $7 + 3 + 4 + 8 + 0 + 0 = 22$
 Difference is 11 divisible by 11.

-by 12 If a number is divisible by 4 as well as 3. It is divisible by 12 i.e The number formed by the last two digits should be divisible by 4; the sum of the digits should be divisible by 3. For eg: 7593684: 84 is divisible by 4; also sum of digits = 42 is divisible by 3; so the no is divisible by 12

-by 14 If the last digit an even number (2, 4, 6, 8, or 0); & if it is divisible by 7. (divide actually by 7 & check up), then it is divisible by 14. (or use the test)

Eg: 1008; 8 is even; also 1008 is divisible by 7 (check up); so 1008 is divisible by 14.

$$-2 \times 8 = \left.\begin{array}{r} 100 \\ 16 \\ \hline 84 \end{array}\right\} \text{div. by 7}$$

-by 15 If the number is divisible by 5 & 3 as well i.e last digit should be 0 or 5, sum of digits is to be divisible by 3;

Eg: 65085

Sum of digits = 24 ∴ divisible by 3; last digit is 5. So it is divisible by 15: Quotient = 4339

-by 16 If the number formed by the last four digits is divisible by 16, then the number will be divisible by 16. 3216 is divisible by 16

Eg: 57893216: → Quotient = 3618326

Prime numbers

An integer greater than 1 that has no factors except 1 & itself is called a prime number. 2 is the first prime number and is the only even prime number

No other even number is a prime number other prime numbers upto 100: 2, 3, 5, 7, 11, 13, 17, 19, 23, 29, 31, 37, 41, 43, 47, 53, 59, 61, 67, 71, 73, 79, 83, 89, and 97.

To Check up: Whether a number is prime or not: (A) let the given number be N say 301

(a) Consider a number $> \sqrt{301}$: since $17^2 = 289$ & $18^2 = 324$; $18 > \sqrt{301}$; let 18 be m.

(b) Consider the prime numbers less than this no m. Here we have to consider 2, 3, 5, 7, 11, 13, 17.

(c) Test whether the given no (here 301) is divisible by any of these numbers.

Check: It is not even; ∄ sum of digits 4 : ∄ last digit not 5 or 0 : ∄ ; It is divisible by 7 so the given no is not a prime no; Factors 7 × 43

(B) Consider: 307 is it a prime no or not? $17^2 = 289$; $18^2 = 324$ ∴ $18 > \sqrt{307}$ find whether 307 is divisible by any of the prime nos < 18 i.e 2, 3, 5, 7, 11, 13, 17. It is not even; ∄ sum of digits = 10∄; last digit is 7∄;

It is not divisible by 7. (by actual division check) or divisibility test $30 - 2 \times 7 = 16$ This is not divisible by 7; sum of even digits = 0; sum of odd digits = 10, difference $0 = \sim 10 = 10$;

∴ It is not divisible by 13 (test by division)

It is not divisible by 17 (test by actual division)

Thus 307 is not divisible by any of the prime nos < 18

Hence we conclude 307 is a prime number

Numbers which are not prime no's are called **composite no.s**

Prime factorization (PF) : The P.F of a no is the expression of the no as the product of its prime factors. Ex.: Find the prime factors of 18354 It is even.

∴ 2 is a factor; sum of digits = 21; 3 is a factor; 4 is not factor, ∄, 7, is a factor actual division check.

∴ $18354 = 2 \times 3 \times 7 \times 437$.; or subtract 2×4 i.e 8 from 1835. We get 1827. This is divisible by 7. (giving 216) ∴ 18354 is divisible by 7.

Is 437 prime? $20^2 = 400$;

$$21^2 = 441$$

∴ $21 > \sqrt{437}$. Now check if 437 is divisible by prime nos < 21 ie 2, 3, 5, 7, 11, 13, 17, 19. ∄, ∄, ∄, ∄, 11 (∵ $11 - 3 = 8$); ∄3 actual check by division by 13, division by 17 &

Check; 19 is a factor.

Actually divide by 19 : $437 = 19 \times 23$, also 23 is prime. Thus prime factors of $18354 = 2 \times 3 \times 7 \times 19 \times 23$. These five prime nos

Co primes: Two no's are co primes if their HCF is 1

   Eg: (2, 3), (2,5), (2,7), (3, 5) etc.

If a no is divisible by two numbers a & b & if a & b are co primes, then the given no is divisible by ab.

Eg: Since 19, 23, are co primes; if 18354 is (given to be) divisible by 19 as well as by 23; (check up, It is) it will also be divisible by 19×23 ie 437.

Check: $18354 = 19 \times 966$; also $18354 = 23 \times 798$;

$\rightarrow 18354 = 437 \times 42$

If a, b are not co primes, a number which is divisible by a & b separately, need not be divisible by ab.

Eg: (i) 210 is divisible by 6

   $(210 = 6 \times 35)$

(ii) 210 is divisible by 14

   $(210 = 14 \times 15)$

(iii) 6 & 14 are not co primes. Their HCF is 2

   $6 = 2 \times 3$

   $14 = 2 \times 7$

(iv) 210 is not divisible by 6×14 i.e 84

$$\begin{array}{r} 84)\overline{210}(2.5 \quad 210 = 84 \times 2.5 \\ \underline{168} \\ 420 \\ \underline{420} \\ x \end{array}$$

Order of Arithmetic Operations PEMDAS

P : Parentheses: ( )

E : Exponentiation: a^b

M : Multiplication ⎫

⎬ In order from left to right

D : Division ⎭

A : Addition
S : Subtraction
} In order from left to right

If an expression has parentheses, with in parentheses, work from the inner most out.

Ex: $96 - 7 \times 5 + (11 - 4)^2 \div 7$

Sol.: First perform any operations with in parentheses

$$96 - 7 \times 5 + 7^2 \div 7$$

Next raise to any powers indicated by exponents.

$$96 - 7 \times 5 + 49 \div 7$$

Then do all multiplication & division in order from left to right

$$96 - 35 + 7$$

Last, do all addition & subtraction in order from left to right.

$$61 + 7 = 68$$

Mnemonic to remember: Please Excuse My Dear Aunt Sally (PEMDAS)

Laws of Operation:

Commutative law Addition (A) & Multiplication (M) are commutative. Subtraction & Division are not. The order is important in them.

Ex: $9 + 8 = 8 + 9; 9 \times 8 = 8 \times 9; 9 - 8 \neq 8 - 9; \dfrac{9}{8} \neq \dfrac{8}{9}$

Associative law: A & M are associative. Terms can be regrouped in any manner.

Ex:
$$\left. \begin{array}{l} (9+8)+7 = 9+(8+7) \\ 17+7 = 9+15 = 24 \end{array} \right\} \begin{array}{l} (9 \times 8) \times 7 = 9 \times (8 \times 7) \\ 72 \times 7 = 9 \times 56 \\ = 504 \end{array}$$

Distributive law: In multiplication a factor can be distributed among the terms being added or subtracted. Division also can be distributed

Numbers 7

$$\left.\begin{array}{r}9(8+7)=9\times 8+9\times 7\\ \text{Ex:} \qquad =72+63\\ =135\end{array}\right\} \quad \begin{array}{c}\dfrac{9-6}{3}=\dfrac{9}{3}-\dfrac{6}{3}\\ =\dfrac{3}{3}\\ =1\end{array}$$

$9(8-7) = 72 - 63 = 9$

How ever when the sum or difference is in the denominator, no distribution is possible.

Ex: $\dfrac{36}{6+3} \neq \dfrac{36}{6} + \dfrac{36}{3} \Rightarrow$ RHS = 4

$\qquad\qquad\qquad\qquad$ RHS = 6 + 12

$\qquad\qquad\qquad\qquad\qquad = 18$

Exercises

[Note : Four or five alternatives are given as possible answers. Indicated by (a), (b), (c), (d), (e). You have to pick up the correct answer.]

1. If a positive number is multiplied by 19 and 23 is added to the product and if the resulting number is divisible by 23 then the least original number is

   (a) 19   (b) 23   (c) 38   (d) 46

2. Thousand times the sum of the digits of a four digit number is subtracted from the number. The resulting number is divisible by

   (a) 1   (b) 3   (c) 9   (d) All these

3. What is the smallest digit to be added to the number 586749, to make the sum divisible by 45?

   (a) 1   (b) 3   (c) 6   (d) 11

4. What is least number to be subtracted from the number 586749 to make the difference divisible by 660?

   (a) 3   (b) 4   (c) 7   (d) 9

5. A number when divided by 118 gives remainder 62. What will be the remainder if the divisor is 59?

118 + 62 = 180
180 ÷ 59 = ?

8   Quantitative Aptitude

(a) 1   (b) 3   (c) 7   (d) 9

6. When a number is divided by 19 the remainder is 17. When the number is divided by 23 the remainder is 15. What is the number?

(a) 245   (b) 307   (c) 423   (d) 518

7. A candidate took 54 as the divisor in place of the given 45 got a quotient of 20 and remainder is 10. What is the correct quotient?

(a) 24   (b) 26   (c) 28   (d) 30

8. When a certain number is divided successively by 6, 7, 8 the remainders are 3, 4 and 6 and the quotient in the last division is 9. What is x?

(a) 3301   (b) 3303   (c) 3506   (d) 3802

9. A number is divided successively by the prime factors (taken in the ascending order) of 231 and the remainders were 2, 2 and 10. What will be the remainder when it is divided by 231?

(a) 212   (b) 215   (c) 218   (d) 223

10. $3^{81} + 3^{82} + 3^{83} + 3^{84}$ is divisible by

(a) 7   (b) 10   (c) 17   (d) 19   (e) 40

| Solutions |

1. $\left[\dfrac{19x+23}{23}\right]$. Let x be the least number

Since both 19, 23 primes, x should be 23.

2. Let the number be (abcd) sum of the digits $= (a+b+c+d)$; If thousand times this is subtracted from (1000a + 100b+10c+d):

We get [1000a + 100b + 10c + d - 1000(a+b+c+d)]

The resulting number will be -900 b-990 c- 999 d.

Numbers 9

This is divisible by 1,3 and 9 also

Ex.: Let the number be 4577 and $4+5+7+7 = 23$.

$$4577$$
$$-230000$$
$$\overline{-225423}$$

$\therefore$ This is divisible by 1,3,9, all

$$5a86$$
$$+9823$$
$$\overline{15b09}$$

3. 586749; To be divisible by 5; 1 or 6 or 11 are to be added

If we add	The Changed number will be	Sum of digits	Is it divisible by 9?	
1	58 67 50	31	X	So add 6 to make it divisible by 5 and 9 and by 45
6	58 67 55	36	✓	
11	58 67 60	32	X	

4. 586749 : Divisible by 660 means; by $3 \times 4 \times 5 \times 11$

So subtract 9; to make the changed number divisible by 3,4,5,11 i.e. 660.

3, 7, 11

$3 \times 2 = 6$ 14

$7 \times 2 = 14$

$11 \times 0 = 110$

If we subtract	Changed number	From right sum of odd digits	Sum of even digits	(Div. by 11): Their difference	(Div. by 3) Sum of all digits	(Div. by 4) Last 2 nos.	Conclusion divisible by
1	586748	23	15	8	38	48✓	4
2	586747	22	15	7	37	47	-
3	586746	21	15	6	36✓	46	2
4	586745	20	15	5	35	45	5✓
5	586744	19	15	4	34	44✓	4
6	586743	18	15	3	33	43	-
7	586742	17	15	2	32	42	2
8	586741	16	15	1	31	41	-
9	586740	15	15	0✓	30✓	40✓	3, 4, 5, 11

Numbers 11

5. $118q + 62 = x \Rightarrow 2 \times 59q + 59 + 3 = x$
   $\Rightarrow (2q+1)59 + 3 = x$ ∴ Remainder will be 3
   check if q is 1; $x = 118 + 62 = 180$.

   ```
   59 ) 180 ( 3
        177
          3
   ```

6. Let q_1, R_1 be the quotient and remainder x be the number.
   $19Q_1 + 17 = x = 23Q_2 + 15$

   Trying the various choices one by one

   (a)
   ```
   19 ) 245 ( 12     23 ) 245 ( 10    So, 245 is one possible
         19                230         answer
         55                 15
         38
         17
   ```

   (b)
   ```
   19 ) 307 ( 16    (c)  19 ) 423 ( 22    (d)  19 ) 518 ( 27
         19                   38                     38
        117                   43                    138
        114                   38                    133
         3x                    5x                    5x
   ```

7. ```
   54 ) x ( 20    ∴ x = 1090;    45 ) 1090 ( 24
       1080                            90
         10                            190
                                       180    Q = 24
                                        10
   ```

8. ```
   6 ) x ( a       7 ) a ( b       8 ) b ( 9
       6a              7b              72
        3               4               6
   ```
   $b = 72 + 6 = 78$; $a = 7b + 4 = 546 + 4$
   $= 550$;   and $x = 6a + 3 = 3300 + 3 = 3303$.

9. Prime factors of 231 are 3, 7, 11

Let n be the number.

```
3)   x
7)  a | 2
12) b | 2
    c | 10
```

If we take conveniently c as 1 then

$b = 11c + 10 = 11 \times 1 + 10 = 21$

$a = 7b + 2 = 7 \times 21 + 2 = 147 + 2 = 149$

$n = 3a + 2 = 3 \times 149 + 2 = 447 + 2 = 449$

If it is directly divided by 231

231) 449 (1   the remainder would be 218
     231

     218

[Note :- If c were to be some other number say 6 then

$b = 11c + 10 = 1 = 11 \times 6 + 10 = 76$

$a = 7b + 2 = 76 \times 7 + 2 = 532 + 2 = 534$

$n = 3a + 2 = 534 \times 3 + 2$

$= 1602 + 2 = 1604$

```
231) 1604 (6
     1386
     ----
      218
```

When this is divided by 231 then the remainder will be 218 (again)

10. $3^{81}(1 + 3 + 3^2 + 3^3) = 3^{81}(1 + 3 + 9 + 27)$

$3^{81} \times 40$ is divisible by 40

Key

1)	b	2)	d	3)	c	4)	d	5)	b
6)	a	7)	a	8)	b	9)	c	10)	e

2

HCF and LCM

Introduction

An integer that is divisible by another integer is a 'multiple' of that integer.

The positive integers that evenly divide into a number are called the 'factors', 'divisors' of that number.

The largest factor shared by two numbers is called the Highest (greatest) Common Factor' of the two (HCF).

The least number divisible by a set of given numbers is called their 'Least Common Multiple' (LCM)

Finding the HCF of two numbers

(a) **Factorization Method:** To find the prime factors of the given numbers, multiply the least powers of the common prime factors to get the HCF.

(1) Ex: HCF of $48, 162$: $48 = 2^4.3$;
$$162 = 3^4.2$$
$\therefore$ HCF $= 2.3$
$= 6$

If there are more than two numbers, find the HCF of two numbers & repeat finding the HCF of this result & the third number & so on.

(2) Ex: HCF of $48, 162, 165$: HCF of $48, 162$ is 6;
HCF of $6, 165$: $6 = 2 \times 3$; $165 = 3 \times 5 \times 11$
$\therefore$ HCF of the three is 3

(b) **Division Method**

Divide the larger number by the smaller. Now divide the first divisor by the remainder in the first division. Repeat the process till we get a zero remainder. The divisor in the last division is the required HCF.

For ex see below (ex 3)

Finding the LCM

(a) Factorization method: To find the prime factors of the given numbers, find the product of the highest powers of all the factors.

For ex see below (ex 4)

(b) Common Division method: Divide the row of the given numbers by numbers which are their common factors carry forward the numbers which are not divisible by them. Repeat the process. Find the product of the successive divisors & the undivided numbers to get the LCM.

For eg see below.

3. Find the HCF of 385, 462 & 1001

(a) Factorization method:

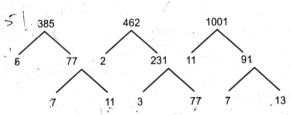

So the no's are $5 \times 7 \times 11, 2 \times 3 \times 7 \times 11, 11 \times 7 \times 13$

Hence HCF is 11×7 or 77

(b) Division Method:

```
385)462(1
    385
    77) 385 (5
        385
         X     HCF of the no's 385, 462 is 77

    77) 1001 (13
        77
        231
        231
         X     HCF of all the three numbers = 77
```

HCF of Decimal fractions their LCM

Change the decimal fractions into those having the same number of decimal places; Find the HCF (or LCM) of the numbers without decimal places; Thereafter mark off decimal places as were in the given numbers.

4. Find the (a) HCF (b) LCM of 0.44, 1.21 & 13.2

Sol.:

(a) First we will rewrite all the numbers given with two decimal places.

0.44, 1.21 & 13.20

Now consider 44, 121 & 1320. What is the HCF for these?

So, they are $2^2 \times 11, 11^2$ & $2^3 \times 5 \times 3 \times 11$

Their HCF is, considering the least powers of the common factors: 11

Therefore the HCF of 0.44, 1.21 & 13.2 is 0.11 answer.

Check: $\dfrac{0.44}{0.11} = 4;\quad \dfrac{1.21}{0.11} = \dfrac{121}{11} = 11;$

$\dfrac{13.2}{0.11} = \dfrac{1320}{11} = 120$

(b) Consider 44, 121, 1320. What is the LCM?

```
11 | 44,  121,  1320
 4 |  4,   11,   120
   |  1,   11,    30
```

$\therefore$ LCM $= 11 \times 4 \times 11 \times 30$

$= 121 \times 4 \times 30$

$= 484 \times 30$

$= 14520$

Therefore the LCM of the given decimal fractions is 145.2 ans.

Check: $\dfrac{145.2}{0.44} = \dfrac{14520}{44}$

$\dfrac{1320}{4} = 330$

$\dfrac{145.2}{1.21} = \dfrac{14520}{121}$ 121) 14520(120 ans

Quantitative Aptitude

$$\frac{145.2}{13.2} = \frac{1452}{132}$$

```
      121
      242
      242
        X

132)1452(11    Checked
    132
    132
    132
      X
```

145.2 is the least common multiple. since for ex $\frac{145.2}{2}$
i.e 72.6 is not a multiple 13.2; for ex $\frac{145.2}{3}$ i.e 48.4 is not a multiple of 13.2 etc

Comparison of Fractions:

First find the LCM of the denominations. Then change the fractions into equivalent fractions with this LCM as the denominator. Now compare the numerators & the one with the largest value is the largest fraction & so on - repeat.

5. Put the fractions in the descending order $\frac{3}{5}, \frac{7}{10}, \frac{11}{20}, \frac{71}{100}$

Sol.: LCM of the denominators LCM $5 \times 2 \times 25 = 100$
So the given fractions are

$\frac{60}{100}, \frac{70}{100}, \frac{55}{100}, \frac{71}{100}$

5	5,	10,	20,	100
2	1,	2,	4,	20
2	1,	1,	2,	10
	1,	1,	1,	5

The required order so is

$\frac{71}{100}, \frac{7}{10}, \frac{3}{5}, \frac{11}{20}$

Exercises

1. The numbers 24, 60 and a third one have as their H.C.F 12. Their LCM is 480. What is the third number?

HCF and LCM   17

(a) 36 (b) 48 (c) 96 (d) 120

2. The sum of two numbers is 396. Their HCF is 11? What are the numbers?

   (a) 22, 264 (b) 30, 366 (c) 160, 264 (d) 187, 209

3. The product of two numbers is 82524; their HCF is 23. The number of such pairs is

   (a) 2 (b) 3 (c) 4 (d) 5

4. The LCM of two numbers is 45. The ratio of the numbers is 3 : 5. what is their difference?

   (a) 3 (b) 6 (c) 9 (d) 12

5. The HCF and LCM of two numbers are 896 and 50176. The ratio of the two numbers is 8 : 7. The smaller of the two numbers is

   (a) 384 (b) 1024 (c) 3072 (d) 6272

6. The product of two co-primes with HCF 1 is 177; their LCM should be

   (a) 1 (b) 177 (c) 531 (d) 1062

7. What are the three primes numbers which are factors 385 and whose LCM is 385?

   (a) 2, 5, 7 (b) 2, 5, 11 (c) 3, 7, 11 (d) 5, 7, 11

8. What is/are the highest divisor of the number 385 (Refer the problem 50), which divide without a remainder?

   (a) 35 (b) 55 (c) 77 (d) 385

9. If one of the two numbers whose HCF and LCM are 3 and 45 is a one digit number, what is the other number?

   (a) 12 (b) 15 (c) 18 (d) 29

10. Three computer operators have to complete a project. They set their 'beepers' which will give a beep (a) after 2 hours (b) after 3 hrs. (c) after 4 hrs. and start the project at 6 a.m. At what time will all the beeping devises 'beep' simultaneously?

18   Quantitative Aptitude

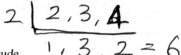

(a) 3 p.m.   (b) 4 p.m.   (c) 5 p.m.   (d) 6 p.m.

11. The sum of two numbers is 759. Their HCF is 23. Which of the following pairs of numbers do(es) not satisfy these conditions

   (a) 23, 736   (b) 46, 713   (c) 230, 529   (d) 229, 530

Solutions

1.

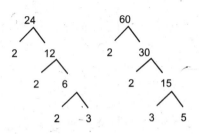

Thus $2^3 \times 3; 2^2 \times 3 \times 5$ and a third number have the HCF of numbers is $2^2 \times 3$

$24 = 2^3 \times 3; 60 = 2^2 \times 3 \times 5; n = ?$

Their

LCM $= 2^5 \times 3 \times 5$

HCF $= 2^2 \times 3$

LCM is $480 = 2 \times 2 \times 2 \times 2 \times 2 \times 3 \times 5$

Third number say is n

$24 = 2^3 \times 3$

$60 = 2^2 \times 3 \times 5$

$= 2^5 \times 3 \times 5$

$n = 25 \times 3 = 32 \times 3 = \mathbf{96}$

Check 24, 60, 96 are divisible by HCF 12,

$\left(\text{we get } \dfrac{24}{12} = 2, \dfrac{60}{12} = 5, \dfrac{96}{12} = 8\right)$

LCM $= 2^5 \times 3 \times 5$, HCF $= 2^2 \times 3$

2. Let the ratio of numbers be a : b. Since HCF is 11

They are 11a, 11b

Their sum is 11a + 11b = 396

∴ a + b = 36.

Out of the given answer sets

   (a) is 22, 264. Their HCF is 11 but their sum ≠ 396×

(b) 30, 366. Their HCF ≠ 11 though their sum = 396 ×
(c) 160, 264. Their HCF ≠ 11. Sum ≠ 396×
(d) 187, 209. Their HCF = 11. Their sum = 396. So (d) is correct

3. The product of two numbers 82524 HCF = 23.

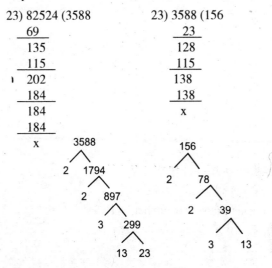

So the factors of 82524 are 23, 23, 2, 2, 3, 13

82524 = 23 × 3588;

$$\left. \begin{array}{l} 3588 = 23 \times 156; \\ = 23 \times 2 \times 2 \times 3 \times 13 \end{array} \right\} \therefore 156 = 12 \times 13 = 2 \times 2 \times 3 \times 13$$

With these factors the several products which can be formed are

	Product	HCF
(23 × 1); (23 × 156)	82524	23 ✓
(23 × 2); (23 × 78) ×	82524	46 x
(23 × 4); (23 × 39)	82524	23 ✓
(23 × 6); (23 × 26) ×	82524	46 x
(23 × 12); (23 × 13)	82524	23 ✓

[These can be the set 3 of factors having HCF as 23]

Out of these: 23 × 2, 23 × 78 is not valid since (2, 78) are not co-primes.

Also (23 × 6), (23, 26) is not valid for the same reason.

Thus $[(23 \times 1), (23 \times 156)]; [23 \times 4, 23 \times 39]; [(23 \times 12), (23 \times 13)]$ Only are the valid pairs i.e. (23, 3588), (92, 897); (276, 299) only are valid pairs.

Check $23 \times 23 = 529; 529 \times 156 = 82524$. Three pairs

4. Let $3x, 5x$ be the number their HCF = x

   LCM is given as 45.

   Product of the numbers $= 3x \cdot 5x = 15x^2$

   HCF $\times$ LCM $= x \times 45$

   $$\therefore 15x^2 = 45x$$
   $$\therefore x = 3$$

   Numbers are $3x, 5x$ i.e = 9, 15. Their difference is 6.

5. Let m, n be the numbers (given) ———— (i)

   Also their HCF is 896, LCM = 50176

   We know product of the two numbers = HCF $\times$ LCM

   $\therefore mn = 896 \times 50176$ ———— (ii)

   From (i) & (ii)

   $$\therefore (mn)\left(\frac{m}{n}\right) = \left(896 \times \overset{7168}{\cancel{50176}}\right) \times \left(\frac{8}{\cancel{7}}\right) = 896 \times 8 \times 7168$$

   $$= 7168^2$$

   $\therefore m = 7168$

   and $n = \dfrac{7m}{8} = \dfrac{7}{\cancel{8}} \times \overset{896}{\cancel{7168}} = 6272$

   Check : Product of $7168 \times 6272 = 44957696 = 896 \times 50176$

   Ratio of 7168 and 6272 = 8: 7 = 1.14286

   Smaller of the number = 6272.

6. HCF of two coprimes = 1

   LCM = 177 (since HCF $\times$ LCM = product of two numbers which is given as 177)

7. They are 5, 7 and 11 as listed out in 50(a) problem.

   LCM $= 5 \times 7 \times 11 = 385$

HCF and LCM 21

8. Divisiors are 1, 5, 7, 11, $5 \times 7, 5 \times 11, 7 \times 11$ and 385. Highest 385

9. Let 3a, 3b be the numbers (since HCF is 3)
   LCM = 9ab = 45 also HCF $\times$ LCM = $3 \times 45 = 3a \times 3b$
   $\therefore$ ab = $\dfrac{45}{3}$ = 15 no's are $3 \times 3, 3 \times 5$ i.e 9, 15

   Since, 3, 6,9 are the only numbers divisible by 3

   '9' is the one digit number, the other number is 15 (since (3×15), (9×5) are the only pairs to give possibly 45 as (Since 5 is not a multiple of 3) LCM)

10. LCM of $\begin{array}{c} 2\lfloor 2, 3, 4 \\ \hline 1, 3, 2 \end{array}$ LCM =12

    12 hrs. ie 6 pm

11. Sum = 759, HCF = 23
    Let the number be 23a, 23b.

    23) 759 (33
        69
        ——
        69
        69
        ——
         0

    $23(a + b) = 759$ $\therefore a + b = 33$
    Only choice 2, 31 $\therefore 2 \times 23 = 46$
                         $31 \times 23 = 713$
                         ———
                         759

    Then the numbers are **46, 713**
    First answer choice : 1×23 and $\dfrac{23 \times 32}{46}$
    First answer choice satisfies $\dfrac{69}{736}$

Possible values	
a	b=(33-a)
1	32
2	31
3	30
4	29
5	28
6	27
7	26
8	25
9	24
10	23
11	22
12	21
13	20
14	19
15	18

2nd answer choice is

Their HCF is 23, sum 759. Second choice satisfies the condition

3rd answer choice is their sum = 759; HCF = 23

$529 = 23^2; 230$. Third answer choice satisfies the condition

4th answer choice 229 not divisible by 23; (which is to be HCF)

Thus 4th choice does not satisfy the condition
(Since also 530 is not divisible by it)

Key

1)	c	2)	d	3)	b	4)	b	5)	d	6)	b
7)	d	8)	d	9)	b	10)	d	11)	d		

3

Fractions: Decimal Fractions; Simplification Problems

Fractions

$\dfrac{a}{b}$: a: numerator (N), b: denominator (D), (-) bar divided by (fraction bar)

Equivalent fractions (EF): When N & D are multiplied by the same no (other than zero) we get an EF

$$\text{Eg: } \frac{1}{3} = \frac{4}{12}$$

When N & D are divided by the same non zero no

We get an E.F

$$\text{Eg: } \frac{6}{8} = \frac{1}{3}$$

Operations with fractions: Addition & Subtraction

Find the LCM of the denominators & add or subtract

$$\frac{3}{5} + \frac{4}{7} - \frac{2}{3} = \frac{63 + 60 - 70}{105}$$

$$\left(\text{since } \frac{63}{105} = \frac{3}{5} \text{ etc}\right)$$

$$= \frac{53}{105}$$

Multiplication & Division

Cancel the common multiples & multiply: Division by a fraction is multiplying by the reciprocal of the dividing factor.

$$\text{Eg: } \frac{5}{3} \div \frac{25}{9} = \frac{5}{3} \times \frac{9}{25} = \frac{3}{5}$$

Comparing fractions. Find the LCM of the denominators. The fraction with largest numerator will be largest.

$$\text{Eg: } \frac{3}{7}, \frac{4}{5} : \frac{15; 28}{35}$$

$$\therefore \frac{4}{5} \text{ is larger}$$

Mixed numbers: Eg: $6\frac{1}{2} = \frac{13}{2}$

Decimal fractions

They are another way of expressing common fractions (CF). They can be converted to CF with a power of ten in the denominator.

$$\text{For eg: } 0.49 = \frac{49}{100}$$

Comparison of decimal fractions

Equalize the number of the digits (by putting zero's to the right if necessary) after the decimal point and compare **operation with decimal fractions (DF) operations** with D.F.

Addition / Subtraction

Line up the decimal points one under the other and perform the operation

Multiplication & Division

Perform the operation. The number of decimal places in the product will be the total number of decimal places in the factors

Ex.: 0.329×0.23

$$0.329$$
$$\underline{0.23}$$
$$987$$
$$\underline{658}$$
$$0.07567 \text{ (no of decimal places} = 3 + 2 = 5)$$

Division: - When dividing a decimal by a decimal multiply each by a power of ten (10) such that the division becomes an integer and carry out the division

Eg.: $0.198 \div 0.33 = \dfrac{19.8}{33}$

$$33\overline{)19.8}(0.6$$
$$\underline{198}$$
$$X$$

Recurring decimal (R.D): A decimal fraction in which, a digit or a set of digits is repeated is called an R.D

Ex.: $\dfrac{1}{6} = 0.166\dot{6}$; $\dfrac{1}{7} = 0.\overline{142857}\,\overline{142857}......$

Pure RD: A decimal fraction in which all the figures after the decimal are repeated like $\dfrac{1}{7}$

Conversion of a pure recurring decimal into a simple fraction

Let $x = 0.333$; $\therefore 10x = 3.333$ subtracting the first

$\therefore 9x = 3$; $\therefore x = \dfrac{3}{9} = 0.333$; Thus $0.333 = \dfrac{3}{9}$

$= \dfrac{1}{3}$

Mixed R.D:- A decimal fraction in which some digits do not repeat & some others repeat **conversion of a mixed RD into a simple fraction.**

Eg: Let $x = 0.246$:

$$1000x = 246.6$$
Then $\quad \underline{100x = 24.6}$
$$900x = 222$$

$\therefore x = \dfrac{222}{900} = \dfrac{111}{450}$

Note: $\overline{4.82}$ **stands for -** $4 + 0.82$

Ex.: 1 Let $x = 26\overline{43}$

$$10000x = 2643.\overline{43}$$
$$\underline{100x = 26.\overline{43}}$$
$$9900x = 2617.0$$

$\therefore x = \dfrac{2617}{9900}$

26 Quantitative Aptitude

Ex.: 2 $4.\overline{97} - 2.\overline{62} = ?$

Sol.: Let $x = 4.\overline{97}$

$$100x = 497.\overline{97}$$
$$\underline{x = 4.\overline{97}}$$
$$99x = 493$$

Let $y = 2.\overline{62}$ | $100y = 262.\overline{62}$
$\underline{y = 2.\overline{62}}$
$99y = 260$

$\therefore y = \dfrac{260}{99}$

$\therefore (x - y) = \dfrac{493}{99} - \dfrac{260}{99}$

$= \dfrac{233}{99}$

$= 2.\overline{35}$ | (also directly) $\begin{array}{r} 4.\overline{27} \\ -2.\overline{62} \\ \hline 2.\overline{35} \end{array}$

Ex.: 3 Find $4.\overline{32} \times 8.\overline{6}$

Sol.: Let $x = 4.\overline{32}$

$$100x = 432.\overline{32}$$
$$\underline{x = 4.\overline{32}}$$
$$99x = 428$$

Let $y = 8.\overline{6}$

$$10y = 86.\overline{6}$$
$$\underline{y = 8.\overline{6}}$$
$$9y = 78$$

$\therefore xy = \dfrac{428}{99} \times \dfrac{78}{9}$

$= \dfrac{33384}{891} = 37.468001$

Ex.: 4 Find $\overline{4}.58 + \overline{3}.31$

Sol.: Given $\Rightarrow -4 + 0.58 - 3 + 0.31$
$= -7.89$
$= \overline{7}.89$

Ex.: 5 Find $\bar{4}.58 + \bar{3}.91$

Sol.: Given $\Rightarrow -4 + 0.58 - 3 + 0.91$
$= -7 + 1.49$
$= -7 + 1 + 0.49$
$= -6.49$
$= \bar{6}.49$ ans

Number properties

Absolute value of a number is the number without its sign
(Its distance from zero on the number line)

Ex.: $191 = 1 - 91$
$= 9$

Properties of zero: When zero is added or subtracted from a number, the number is not changed

$$16 + 0 = 16$$
$$28 - 0 = 28$$

When a no is multiplied by zero, it becomes 0: $27 \times 0 = 0$
Division by zero is not defined;
$\dfrac{0}{0}$ is also not defined.

Properties of 1 Any number when multiplied by 1 will be the same. When divided by 1 also remains same

Ex.: $72 \times 1 = 72$; $\dfrac{63}{1} = 63$; $-84 \times 1 = -84$;

$$\left(\dfrac{-47}{1}\right) = -47$$

Properties of (-1) When a number is multiplied or divided by (-1), it changes its sign. $4(-1) = -4$; $\dfrac{17}{(-1)} = -17$; $\dfrac{(-3)}{(-1)} = 3$; $(-2)(-1) = +2$

Operations with signed numbers

Addition: $3 + 5 = 8$; $(-3) + (-2) = (-5)$ like signs; unlike signs: $(-6) + (2) = (-4)$

Quantitative Aptitude

(sign of the no with larger abs value)

Subtraction:
$$4 - 2 = 2; (-4) - (-2) = -4 + 2$$
$$= (-2)$$
$$-4 - 2 = -6;$$

Multiplication:
$$(-2) \times (-3) = 6$$
$$(-2) \times (3) = -6$$

Odd & even:- Numbers divisible by 2 are even otherwise odd; 0 is even
Rules for odd & even numbers
Odd + Odd = Even; Odd × Even = Odd
Even + Even = Even; Even × Even = Even
Odd + Even = Odd; Odd × Even = Even

Exercises

1. Express as a simple fraction 0.00125 of 24/5
   (a) 6×10^{-3} (b) 6×10^{-2} (c) 6×10^{-1} (d) 6

2. Find the approximate value of $25 \times 0.66 \times 0.50 \times 0.19 \times 0.625 \times 0.75$
   (a) 0.27   (b) 0.78   (c) 0.87   (d) 0.93

3. If a square yard plot near a village is sold at Rs. 950/-. What is the cost of 10 acres of a land? (1 acre = 4840 square yards)
   (a) Rs. 45980/-   (c) Rs. 4598×10^3/-
   (b) Rs. 459800/-   (d) Rs. 4598×10^4/-

4. The value of $0.\bar{4} + 0.\bar{5} + 0.\bar{6} + 0.\overline{32}$ is
   (a) $\dfrac{19.7}{9}$ (b) $\dfrac{197}{99}$ (c) $\dfrac{197}{100}$ (d) $\dfrac{1970}{999}$

5. Given $4.44 \times 99 = 439.56$. What is 0.99×444000?

(a) 43.956×10^3 (c) 4395.6×10^3
(b) 439.56×10^3 (d) 43956×10^3

6. When 25740 is divided by 1716 the quotient is 15. What is the quotient when 25.74 is divided by 1500?

    (a) 0.01716   (b) 0.1716   (c) 1.716   (d) 17.16

7. Find the value of $1 + \dfrac{1}{3} + \dfrac{1}{3^2} + \dfrac{1}{3^3} + \ldots\ldots + \dfrac{1}{3^{10}}$

    (a) 1.489   (b) 1.499   (c) 1.509   (d) 1.519

8. $\dfrac{2.98 \times 5.003 \times 29.96}{16.001 \times 20.001}$ is closest to

    (a) 1.4   (b) 14   (c) 1.8   (d) 2.4

9. $\dfrac{(0.03)(0.03)(0.03) - (0.01)(0.01)(0.01)}{(0.03)(0.03) - (0.01)(0.01)} = ?$

    (a) 0.1   (b) 0.3   (c) 1   (d) $\dfrac{13}{400}$

10. $16^4 - 14^4 = ?$

    (a) 3390   (b) 6780   (c) 27120   (d) 54240

Solutions

1. $\dfrac{125}{100000} \times \dfrac{24}{5} = \dfrac{2400}{4 \times 10^5} = \dfrac{600}{10^5} = \dfrac{6}{10^3} = 6 \times 10^{-3}$

2. $\dfrac{100}{4} \times \dfrac{2}{3} \times \dfrac{1}{2} \times \dfrac{2}{10} \times \dfrac{625}{1000} \times \dfrac{3}{4}$

$= \dfrac{625}{800} = \dfrac{25}{8 \times 4} = \dfrac{25}{32} = 0.78 = 0.8$

```
32)250(0.78
   224
   260
   256
     4
```

3. 10 acres = 48400 sq. yards

   1 sq. yard — Rs. 950/-

   ∴ 48400 sqd cost → 48400 × 950

   = Rs. 4598×10^4/-

4. $0.\bar{4} = \dfrac{4}{9}$ check: if $n = 0.4\bar{4}$ then

   $$\begin{array}{rl} 10n = & 4.\bar{4} \\ n = & 0.\bar{4} \\ \hline 9n = & 4 \\ n = & 4/9 \end{array}$$

   $0.\bar{5}$ let $n = 0.5\bar{5}$ then

   $$\begin{array}{rl} 10n = & 5.\bar{5} \\ n = & 0.\bar{5} \\ \hline 9n = & 5 \\ n = & 5/9 \end{array}$$

   $0.\bar{6}$ if $n = 0.6\bar{6}$ then

   $$\begin{array}{rl} 10n = & 6.\bar{6} \\ -n = & 0.\bar{6} \\ \hline \text{Subtracting:} \quad 9n = & 6 \\ n = & 6/9 \end{array}$$

   Let $n = 0.32\overline{32}$ then

   $$\begin{array}{rl} 100n = & 32.32\overline{32} \\ n = & 0.32\overline{32} \\ \hline 99n = & 32 \\ n = & 32/99 \end{array}$$

   Consider $\dfrac{4}{9} + \dfrac{5}{9} + \dfrac{6}{9} + \dfrac{32}{99} = \dfrac{44 + 55 + 66 + 32}{99} = \dfrac{197}{99}$

5. $4.44 \times 99 = 439.56$; ∵ $0.99 = 99 \times 10^{-2}$

   $444000 = 4.44 \times 10^5$

   Thus $444000 \times 0.99 = 4.44 \times 10^5 \times 99 \times 10^{-2} = 439.56 \times 10^3$

6. $\dfrac{25740}{1716} = 15 = \dfrac{25740}{15} = 1716$

Now $25.74 \Rightarrow = 25.740$

$\qquad\qquad\qquad = 25740 \times 10^{-3}$ and $1500 = 15 \times 10^2$

$\therefore \dfrac{25740 \times 10^{-3}}{15 \times 10^2} = 1716 \times \dfrac{10^{-3}}{10^2}$

$\qquad\qquad\qquad = 1716 \times 10^{-5} = 0.01716$

7. $a_1 = \dfrac{1}{3^0};\ CR = \dfrac{1}{3};\ T_{11} = a_1 r^{10}$ and sum of 11 terms

$$\Rightarrow S_{11} = \dfrac{a(1-r^n)}{(1-r)} = \dfrac{1}{3^0} \dfrac{\left[1 - \left(\dfrac{1}{3}\right)^n\right]}{\left[1 - \dfrac{1}{3}\right]}$$

$$= \dfrac{1 - \dfrac{1}{3^n}}{\dfrac{2}{3}}$$

$$= \dfrac{1 - \dfrac{1}{3^{11}}}{\dfrac{2}{3}} = \dfrac{3^{11} - 1}{2 \times 3^{10}} = \dfrac{(81 \times 81 \times 27 - 1)}{2 \times 81 \times 81 \times 9}$$

$$= \dfrac{177146}{118098} = 1.499$$

Check:

sum $= 1 + \dfrac{1}{3} + \dfrac{1}{9} + \dfrac{1}{29} + \dfrac{1}{81} + \dfrac{1}{243} + \dfrac{1}{729} + \dfrac{1}{2187} + \dfrac{1}{6561} + \dfrac{1}{19683}$

$= \dfrac{19683 + 6561 + 2187 + 729 + 243 + 81 + 27 + 9 + 3 + 1}{19683}$

$= \dfrac{29524}{19683}$

Now $\dfrac{177146}{118098} = \dfrac{29524.3}{19683} \simeq \dfrac{29524}{19683} = 1.499$ QED

8. $\simeq \dfrac{3 \times 5 \times 30}{16 \times 20} = \dfrac{45}{32} = 1.406 \simeq 1.4$

9. Let $a = 0.03$, $b = 0.01$, Given $\dfrac{a^3 - b^3}{a^2 - b^2}$

$$= \dfrac{(a-b)(a^2 + ab + b^2)}{(a-b)(a+b)} = \dfrac{(0.03^2 + 0.0003 + 0.0001)}{0.03 + 0.01}$$

$$= \dfrac{0.0009 + 0.0003 + 0.0001}{0.04} = \dfrac{0.0013}{0.04} = \dfrac{13}{400}$$

10. Let $a = 16$, $b = 14$

Given $a^4 - b^4 = (a^2 + b^2)(a^2 - b^2)$

$= (a^2 + b^2)(a+b)(a-b) = (16^2 + 14^2)(16+14)(16-14)$

$= (256 + 196)(30)\,2 = 452 \times 30 \times 2 = 13560 \times 2$

$= 27120$

Key

1)	a	2)	b	3)	d	4)	b	5)	b
6)	a	7)	b	8)	a	9)	d	10)	c

4

Simplification

Exercises

1. The highest and lowest percentage marks are 89 & 12. If they are converted to 20. What is the difference of the marks?

   (a) 11.5   (b) 12.3   (c) 15.4   (d) 208

2. One fifth a person's savings in National Savings Certificate (NSC) is equal to one fourth of his savings in public provident fund. If his total savings are Rs.140000/-. What are the savings in NSC?

   (a) 12673.5   (b) 13982.1   (c) 146732.2   (d) 15555.6

3. If an amount of Rs.4000/- is divided among x, y and z; if x gets 7/8 of what y gets and y gets 1/5th of what z gets; what is y's share?

   (a) Rs.145.5/-   (c) Rs.581.8/-
   (b) Rs.290.9/-   (d) Rs.1163.6/-

4. Water which can fill 3 large bottles can fill 8 small bottles. What is the fraction which can fill a small bottle compared to that in a large bottle?

   (a) 3/8   (b) 1/2   (c) 5/8   (d) 3/4

5. 40 buckets of water fill a tank fully. If a new bucket of 2/5th of the capacity of the old one is used, how many new buckets will fill the same tank?

(a) 70   (b) 80   (c) 90   (d) 100

6. A person gives half the money to his wife in the will, one third of the remaining amount to his two sons and the balance to his two daughters. If each daughter gets Rs.30,000/-. What is the amount he gave to his two sons?

    (a) Rs.15,000/-   (b) Rs.20,000/-   (c) Rs.25,000/-   (d) Rs.30,000/-

7. At an international party 1/4 of the participants were Pakistani men. If the Pakistani men were 2/3 greater than pak women at the party what fraction of the participants were non pakistanies.

    (a) 1/5   (b) 2/5   (c) 3/5   (d) 4/5

8. 'A' throws the dart 4 times for every three throws of 'B'. 'A' hits the target once in 3 throws; while B hits once in two throws. If B has missed 18 times, how many times A has hit the target?

    (a) 4   (b) 8   (c) 16   (d) 32

9. $x - y = 7$, $x^2 + y^2 = 85$. Find xy

    (a) 2   (b) 9   (c) 18   (d) 81

10. 'A' gets in to an elevator at the 4th floor and goes up at the rate of 9 floors per minute. At the same time B starts at the 24th floor and comes down at the rate of 11 floors per minute. At which floor will they cross each other?

    (a) 7   (b) 9   (c) 11   (d) 13

11. A group of persons wanted to raise a contribution of 1 lakh. Had they contributed Rs.50/- more each, the contribution would have been more by Rs.25000/-. How many persons are there?

    (a) 200   (b) 300   (c) 400   (d) 500

12. Note books (at the rate of 1/10th of the No. of children each) were distributed to a class. If the strength of the children is 2/3 of the actual, each would have got 9 books. How many books were distributed?

    (a) 6 (b) 40 (c) 60 (d) 360

13. In a class 1/3 of the students learnt Hindi also in addition to the second language Telugu) and 1/2 learnt Sanskrit. If 1/5 of the students know both Hindi and Sanskrit, what the fraction who do not know either language?

    (a) 11/30 (b) 11/15 (c) 11/5 (d) 11/3

14. For a 30 days, 4 Sunday month, for a group of 10 daily wage earners whose daily earning is Rs.80/- what is the earning if due to a cyclone they could not go for work for 3 days?

    (a) Rs.2,300/-   (b) Rs.4,600/-
    (c) Rs.9,200/-   (d) Rs.18,400/-

15. In a hotel 70% had vegetarian snacks while 20% had non - vegetarian items; if 16% had both types of snacks, how many did not eat either type of snacks if total number of people present were 50?

    (a) 7 (b) 9 (c) 13 (d) 15

Solutions

1. $89\% = \dfrac{17.8}{20}$ and $12\% = \dfrac{2.4}{20}$

   $\therefore$ difference $= 17.8 - 2.4 = 15.4$

2. Let N be savings in NSC; 140,000 - N in PPF say P; $\dfrac{N}{S} = \dfrac{14,0000 - N}{4}$

   $\Rightarrow$ $4N = 700,000 - 5N$ or $9N = 700,000$

   $\therefore$ $N = 77,777.8$; $P = 6222.2$

   Check $\dfrac{N}{5} = 15,555.6$ $\dfrac{P}{4} = 15,555.7$ QED

Quantitative Aptitude

3. $x = \frac{7}{8}y$; $y = \frac{1}{5}z$; $x + y = z = 4000$

$$\therefore x = \frac{7}{8} \times \frac{1}{5}z = \frac{7}{40}z$$

$$\therefore \frac{7}{40}z + \frac{1}{5}z + z = 4000 \Rightarrow \frac{7z + 8z + 40z}{40} = \frac{55}{40}z$$

$$\therefore z = \frac{4000 \times 40}{55} = \frac{32000}{11} = 2,909$$

$$y = 581.8, \quad x = \frac{4072.6}{8} = 509.075 \to x$$

$$x + y + z = \begin{array}{l} \text{Rs.}581.8 \to y \\ 2909 \to z \\ \overline{3999.875} \end{array} \quad \text{Total } 4000/\text{-Rs}$$

4. 3 large bottles = 8 small bottles

   1 small bottle = 3/8 of large bottle

5. Let b_1 be the capacity of a bucket = 1/40 of tank; new bucket $b_2 = \frac{2}{5}b_1$ Full tank = 40 bottles

$$b_2 = \left(\frac{2}{5}b_1\right)$$

   $\therefore$ Full tank = $40\, b_1$;

   $\therefore$ Full tank = $40 \cdot \frac{5}{2} b_2 = 100\, b_2$

6. $\frac{x}{2} + \frac{1}{3}\left(\frac{x}{2}\right)$ he gave to his wife and two sons of out of x

$$x - \frac{x}{2} - \frac{x}{6} = \frac{6x - 3x - x}{6} = \frac{2x}{6} = \frac{x}{3}$$

   Each daughter gets $\frac{1}{2} \times \frac{x}{3} = \frac{x}{6} = 30,000$

   $$\therefore x = 1,80,000$$

   Each son gets $\frac{1}{2} \times \frac{1}{3} \times \frac{1,80,000}{2} = \text{Rs.}15,000$

$$\begin{aligned}
\text{Wife:} &\quad 90{,}000\\
\text{2 sons:} &\quad 30{,}000\\
\text{2 daughters:} &\quad \underline{60{,}000}\\
\text{Total:} &\quad 1{,}80{,}000
\end{aligned}$$

7. Let Pakistani men be M, pak women be w at the party

Total participants = P;

$$\frac{P}{4} = M;\ M - W = \frac{2}{3}W \therefore M = \frac{5}{3}W$$

$$\therefore M + W = \frac{5}{3}W + W = \frac{8W}{3} \text{ and}$$

$P = 4M$

$$\therefore M = \frac{P}{4};\ W = \frac{3}{5}M = \frac{3}{5} \times \frac{P}{4} = \frac{3P}{20}$$

$$\therefore M + W = \frac{8}{3} \times W = \frac{8}{3}\cdot\frac{3P}{20} = \frac{8P}{20}$$

$$\text{Non Pakistanies} = P - \frac{8P}{20} = \frac{12P}{20} = \frac{3P}{5}$$

Required fraction = 3/5

Check If $P = 500$; $M = 125$; $W = \frac{3}{5}M = \frac{3}{5} \times 125 = 75$

$M + W = 200$; Non Pakistanis = 300

$$\text{Fraction} = \frac{3}{5} \text{ QED}$$

8. Let $4N$ and $3N$ be the number of throws by A and B respectively. A hits the target $\frac{4N}{3}$ times B hits is $\frac{3N}{2}$ and misses $\frac{3N}{2}$ times, $\frac{3N}{2}$ is given as 18

$$\therefore N = \frac{2 \times 18}{3} = 12 \quad \therefore 4N = 48$$

$3N = 36$ is the number of trials by A and B.

A hits $\frac{4N}{3}$ (or) $\frac{48}{3} = 16$ times

Check A throws 48 times, hits $\frac{48}{3} = 16$ times B throws 36 times misses 18 times total throws = 84

9. $(x-y) = 7$; $x^2 + y^2 - 2xy = 49$; $x^2 + y^2 = 85$;
   Subtracting $2xy = 85 - 49 = 36$

   $$\therefore xy = 18$$

   [Since the factors are $[18 \times 1, 9 \times 2, 6 \times 3]$
   x and y could be 9 and 2 respectively [or] the other way

   $$x^2 + y^2 = 85, \ x - y = 7$$

   Only $x = 9$, $y = 2$ [To satisfy $x - y = 2$]

10. Let them meet after t minutes

    $$4 + 9t = 24 - 11t$$
    $$\therefore 20t = 20$$

    $t = 1$ min at the 13th floor

11. Let the contribution of each be 'c' rupees and let 'n' be no. of persons
    'nc' = 1,00,000 also $n(c + 50) = 1,25,000$
    Subtracting $50n = 25,000$

    $\boxed{n = 500}$;   c = 200

12. Let 'n' be no. of children. B be the no. of Books each got n/10 = B;
    n = 10B also $\left(\dfrac{2}{3} n\right) 9 = n \dfrac{n}{10}$

    $$\therefore 6n = \dfrac{n^2}{10}$$
    $$n = 60$$

    Check 60 children each got $\dfrac{60}{10} = 6$

    Total Books 360 if children are $\dfrac{2}{3} (60) = 40$

    Each getting 9. Total = 360

13. Let x be the required fraction

    Then $\left(\dfrac{1}{3} - \dfrac{1}{5}\right) + \dfrac{1}{5} + \left(\dfrac{1}{2} - \dfrac{1}{5}\right) + x = 1$

    or $\dfrac{10 - 6 + 6 + 15 - 6 + 30x}{30} = 1$

∴ $10 + 15 - 6 + 30x = 30$
∴ $30x = 30 - 25 + 6 = 11$
∴ $x = 11/30$

Check If 60 is the total number of students

Only Hindi = $20 - 12 = 8$
Only Sanskrit $30 - 12 = 18$

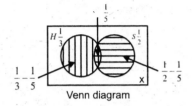
Venn diagram

Both H & S	→ 12
None of the two	→ 22
Total	60

QED

14. Normal working days in the month = 26; cyclone 3 days

∴ Actual working days = $26 - 3 = 23$
Earning = 23 days × Rs. (80/day) × 10 workers
= 230×80 = Rs.18,400

15. Total = 50; veg → 70% = 35;
Non veg = $\frac{1}{5} \times 50 = 10$
Both = $\frac{16}{100} \times 50 = 8$
None = $50 - (27 + 8 + 2) = 50 - 37 = 13$

Key

1)	c	2)	d	3)	c	4)	a	5)	d
6)	d	7)	c	8)	c	9)	c	10)	d
11)	d	12)	d	13)	a	14)	d	15)	c

5

Powers and Roots

Introduction

Powers

In $7a^2$, 7 is the coefficient; a is the base, & 2 is the exponent (The number of times the base is multiplied by itself)

To multiply terms with same base; keep the same base & add the exponents

$$\text{Eg:} \quad 7^3 \times 7^5 = 7^{3+5}$$
$$= 7^8$$

To divide: Subtract the exponent of the denominator from that of the numerator.

$$\text{Eg:} \quad \frac{7^{10}}{7^3} = 7^{10-3}$$
$$= 7^7$$

To raise a power to another power: multiply the exponents:

$$(7^3)^4 = 7^{3 \times 4} = 7^{12}$$

Any number (except zero) raised to the zero power is equal to 1.

Eg: $a^0 = 1$; $(a \neq 0)$:
Eg: $7^0 = 1$
0^0 is not defined

A negative exponent indicates a reciprocal: $7^{-3} = \dfrac{1}{7^3}$

Roots: A fractional exponent indicates a root: $a^{\frac{1}{n}} = \sqrt[n]{a}$

If no 'n' is present, the radical sign simply means a square root: $\sqrt{7} = 7^{1/2}$;

$$8^{-1/2} = \frac{1}{\sqrt{8}} = \frac{1}{8^{1/2}}$$

Every positive number has two square
Roots: One positive, the other negative:
Eg: $\sqrt{49} = +7$ or -7

Rules of Operation of Roots

By **Convention** the **radical** (the symbol $\sqrt{}$) means the **positive** square root only

Addition & Subtraction: Only like radicals can be added to or subtracted from one another.

Eg: $4\sqrt{5} + 3\sqrt{7} - 2\sqrt{5} + \sqrt{7} = 2\sqrt{5} + 4\sqrt{7}$

Multiplication & Division: To multiply or divide one radical by another; multiply or divide the coefficients first, then the radicands

Eg: (1) $2\sqrt{7} \times 5\sqrt{5} = 10\sqrt{35}$

Eg: (2) $\dfrac{10\sqrt{12}}{2\sqrt{4}} = \dfrac{10}{2}\sqrt{\dfrac{12}{4}} = 5\sqrt{3}$;

Eg: (3) simplification: $\sqrt{12} = 2\sqrt{3}$

Powers of 10: Multiplication: If a number is multiplied by a power of 10, move the decimal point to the right the same number of places as the exponent of 10

Eg: $7 \times 10^3 = 7000$;

Eg: $7.23 \times 10^3 = 7230$

Division: Here move the decimal to the left

Eg: $7.23 \times 10^{-3} = 0.00723$

Scientific Notation: is expressing a number as the product of a decimal between 1 & 10, & a power of 10

Large numbers or small decimal fractions are conveniently expressed in this notation.

Eg: 1) $2300000 = 2.3 \times 10^6$ (2.3 million)

Eg: 2) $0.000049 = 4.9 \times 10^{-5}$

Eg: 3) $0.0054 \times 10020 = 5.4 \times 10^{-3} \times 1.002 \times 10^4$

$$= 5.4108 \times 10 = 54.108$$

$$= 5.4108 \times 10^1$$

Exercises

1. The least perfect square which is divisible by 63, 77, 99, 143 is

   (a) 117117 (b) 234234 (c) 585585 (d) 9018009

2. What is the least number to be subtracted from 0.000843 to make it a perfect square?

   (a) 1.5×10^{-6}
   (b) 1.8×10^{-6}
   (c) 2×10^{-6}
   (d) 2.2×10^{-6}

3. What is the smallest number by which 10647 is to be multiplied to make it a perfect square?

   (a) 3 (b) 7 (c) 11 (d) 13

4. What is the smallest number larger than 10000 which is a perfect square?

   (a) 10009 (b) 10049 (c) 10064 (d) 10201

5. What is the perfect cube very close to and higher than the smallest four digit number?

   (a) 1027 (b) 1164 (c) 1208 (d) 1331

6. With what number should we multiply 72000 to make it a perfect cube?

   (a) 1 (b) 2 (c) 3 (d) 4

7. Which number is to be added to 21 to make it a perfect cube?

   (a) 1 (b) 2 (c) 4 (d) 6

8. What number should be divided by $\sqrt{225}$ to give 12?

   (a) 45 (b) 90 (c) 180 (d) 360

9. $(676)\sqrt{x} + 728 = \dfrac{12}{13} \times 1521$. What is x?

   (a) 1   (b) 2   (c) 3   (d) 4

10. Given $2\sqrt{5} = 4.47$.
    Find the value of $4.47 + \sqrt{8000} + 3\sqrt{320} + \sqrt{980}$

    (a) 22.35   (b) 44.7   (c) 89.4   (d) 178.8

11. Given $a = 0.1222$. Find $\sqrt{16a^2 - 8a + 1} + 4a$

    (a) 1.0000   (b) 2.2444   (c) 3.3666   (d) 4.4888

12. The least perfect square divisible by 12, 54, 72 is

    (a) 324   (b) 648   (c) 1296   (d) 2592

13. What is the smallest number to be subtracted from 3421 to make it a perfect square?

    (a) 19   (b) 38   (c) 57   (d) 76

14. What is the smallest number by which 98 is to be multiplied to make it a perfect cube?

    (a) 7   (b) 14   (c) 28   (d) 56

Solutions

1. The LCM of 63, 77, 99, 143, is $7 \times 9 \times 11 \times 7$

   This must be multiplied by $7 \times 11 \times 13$

   (Since it is $3^2 \times 7 \times 11 \times 13$) i.e by $7 \times 143 = 1001$

   $\therefore$ Required $9 \times 1001 \times 1001 = 9009 \times 1001 = 9018009$

   Check its square root is **3003**

   Also $\dfrac{9018009}{63} = \dfrac{1002001}{7} = 143143$ (Div. by 63)

   This is also divisible by 11, 13

   $\dfrac{9018009}{77} = \dfrac{819819}{7} = 117117$

= 117117 (Div. by 77)

This is also divisible by 9, 13

$$\frac{9018009}{99} = \frac{819819}{9} = 91091 \text{ (Div. by 9)}$$

This is also divisible by 7, 13

$$\frac{9018009}{143} = \frac{819819}{13} = 63063 \text{ (Div. by 143)}$$

2. Given = 843×10^{-6}
   Subtract 2×10^{-6}
   To make it 841×10^{-6}
   Its square root is 29×10^{-3}

$$\begin{array}{r|l} \overline{8\,4}\,\overline{3} & 29 \\ 4 & \\ \hline 49\,|\,4\,9\,3 & \\ 4\,4\,1 & \\ 2 & \end{array}$$

3. Multiply by 7. Since,

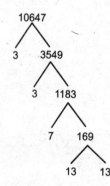

4. Since $\sqrt{10000}$ is 100

   The next square is 101^2

   $$\begin{array}{r} 101 \times 101 \\ \hline 101 \\ 000 \\ 101 \\ \hline 10201 \end{array}$$

5. Smallest four digit number = 1000; $\sqrt[3]{1000} = 10$

   Next higher is $11^3 = 121 \times 11 = 1331$

6. $72000 = 72 \times 1000 = 72(10)^3$

   $72 = 2 \times 2 \times 2 \times 3 \times 3$

   It should be multiplied by 3, we get 21,6000;

   $(60)^3 = 3600 \times 60 = 216000;\ \sqrt[3]{216000} = 60$

7. We see that

   $2^3 = 8,\ 3^3 = 27$, so **6** is to be **added**

8. Given $\dfrac{x}{\sqrt{225}} = \dfrac{x}{15} = 12$

   $\therefore\ x = 180$

9. $676\sqrt{x} = 12 \times 117 - 728 = 1404 - 728 = 676$

   $\therefore\ \sqrt{x} = \dfrac{676}{676} = 1$

   $\therefore\ x = 1$

10. Given

    $= 2\sqrt{5} + \sqrt{1600 \times 5} + 3\sqrt{64 \times 5} + \sqrt{196 \times 5}$

    $= 2\sqrt{5} + 40\sqrt{5} + 3 \times 8\sqrt{5} + 14\sqrt{5}$

    $= 2\sqrt{5}\,[1 + 20 + 12 + 7]$

    $= 2\sqrt{5}\,[40] = 4.47 \times 40 = 178.80$

11. Given $= \sqrt{(4a-1)^2} + 6a$ since $4a < 1$

    $= 1 - 4a + 6a$
    $= 1 + 2a$
    $= 1 + 0.2444 = 1.2444$

12. LCM of 12, 54, 72:

6	12, 54, 72
2	2, 9, 12
3	1, 9,
	1, 3, 2

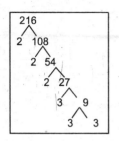

LCM = 216

∴ $216 = 2^3 \times 3^3$

Multiply this by 2×3 to make it a perfect square of $(2^2 \times 3^2)$ i.e 36

∴ Required No = $216 \times 6 = 1296$

13. Subtract 57

To make the number 3421
                     - 57
                    ─────
                     3364

To make it a perfect square = $58 \times 58 = 3364$

14.

This is to be multiplied by $2 \times 2 \times 7$ i.e, 28 to make it 2744 which is a perfect cube

Check:

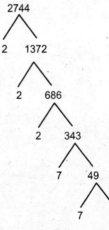

So $2744 = 2^3 \times 7^3 = (2 \times 7)^3$ is a perfect cube

∴ 98 is to be multiplied by **28**

To make the resulting number a perfect cube

Key

1)	d	2)	d	3)	b	4)	d	5)	c	6)	c
7)	d	8)	c	9)	a	10)	d	11)	a	12)	c
13)	c	14)	c								

6

Word Problems on Numbers

Exercises

1. When $\frac{1}{6}$th of a number is added to the number, the sum is 4935. What is the number?

   (a) 1050 (b) 2130 (c) 4230 (d) 8260

2. What is the smallest fraction by which 1485 is to be multiplied so that the resulting number is divisible by 36?

   (a) $\frac{1}{2}$ (b) 1 (c) $\frac{4}{3}$ (d) 2

3. The product of two numbers is 288. The sum of their squares is 580. What is the sum of the numbers?

   (a) 16 (b) 18 (c) 34 (d) 42

4. The difference of two positive integers is 5, the sum of their squares is 377, What is the difference of their squares?

   (a) 121 (b) 135 (c) 156 (d) 352

5. The sum of four consecutive number is 1250. What is the largest?

   (a) 314 (b) 628 (c) 942 (d) 1256

6. The difference of a two digit number and the number obtained by interchanging the two digit is 36. Find the pair if both is prime numbers

   (a) 15, 51   (b) 62, 26   (c) 73, 37   (d) 95, 59

7. By interchanging the digits of a two digit number we get a number which is four times the original number minus 24. What is the original number if the difference of the units digits and tens digit is 7?

   (a) 18   (b) 29   (c) 36   (d) 58

8. The tens place digit is 3 times the units place digit of a two digit number. The difference of the two digit number and the number obtained by interchanging the digits is 54. What is the product of the two digits?

   (a) 9   (b) 18   (c) 27   (d) 54

9. The middle digit of a three digit number is one third of the sum of the other two. The sum of the three digits is 16. The difference of the number obtained by reversing the digits and the original number is 198. What is the number?

   (a) 457   (b) 547   (c) 745   (d) 754

10. In a fraction the difference in the numerator & denominator is 7. When the denominator is increased by 3, the ratio is decreased to 9/10. What is the original fraction?

    (a) 5/12   (b) 7/5   (c) 2   (d) 12/5

11. The numerator & denominator of a fractions are in the ratio of 3/4. When numerator is reduced by 5; the resulting fraction is 1/6th of the original fraction. What is numerator of the original fraction?

    (a) 3   (b) 4   (c) 6   (d) 8

12. If 2 is subtracted from the numerator of a given fraction it becomes 2/3 of the original fraction. Whereas if 3 is subtracted from the denominator it becomes 1/2. What is the original fraction?

    (a) $\dfrac{1}{3}$ (b) $\dfrac{6}{15}$ (c) $\dfrac{7}{15}$ (d) $\dfrac{8}{15}$

13. The sum of the squares of three numbers is 565; The sum of their products taken two a time is 558. Find the sum of the numbers?

    (a) 39   (b) 41   (c) 43   (d) 45

14. When the difference of a two digit number and the number obtained by reversing the digits is found (irrespective of whether the tenths digits is more than units place digit or otherwise in the original numbers); the difference is divisible always by

    (a) 3   (b) 5   (c) 7   (d) 9

15. The difference of a six digit number and the number obtained by reversing the number irrespective of which ever is large is always divisible by

    (a) 3   (b) 9   (c) 90   (d) 99

16. The difference of a seven digit number and the number obtained by reversing the number irrespective of which ever is larger is always divisible by

    (a) 3   (b) 9   (c) 99   (d) 990

17. The sum of four number is 18 if p, q, r, s, are the numbers p - 4; q + 4, 2r and $\dfrac{s}{2}$ are all equal to each other. What is the difference of the largest and smallest numbers?

    (a) 2   (b) 4   (c) 6   (d) 8

Solutions

1. If x is the number

$$x + \frac{x}{6} = \frac{7x}{6} = 4935$$

$$\therefore x = \overset{705}{\cancel{4935}} \times \frac{6}{7}$$

$$= \mathbf{4230} \text{ ans}$$

Check: 4230 + 705 = 4935

2.
```
            1485
           /    \
          3     495
               /   \
              3    165
                  /   \
                 3    55
                     /  \
                    5   11
```

$1485 = 3^3 \times 5 \times 11$ if the resulting number (after multiplying 1485 by a fraction) is to be divisible by 36 and if the fraction is to be smallest we should divide by 3 and multiply by 4 the given number, 1485, so that we get $1485 \times \frac{4}{3} = 495 \times 4 = 1980$. The smallest multiplier fraction is $\frac{4}{3}$ ans

Check: $1980 = \frac{4}{3} \times 1485$; 1980 is divisible by $\frac{3}{4}$

1980 is div. by $3^2 \times 2^2 = 36$

$$\left(\because \left(\frac{4}{3} \times 3^3 \times 5 \times 11 \right) \div \frac{1}{3^2 \times 2^2} = 5 \times 11 \right)$$

Aliter: Let $x = 1485 = 27 \times 55$ ...(i)

If the new number is x_1

We have $\frac{x_1}{36} = 55$

Then $\frac{x_1}{36} = \frac{x}{27}$ from (i)

Word Problems on Numbers   53

$$\therefore x_1 = \frac{36}{27}x = \frac{4}{3}x \text{ QED}$$

$$\therefore x_1 = \frac{4}{3}x \text{ ans}$$

3. $xy = 288, x^2 + y^2 = 580$

   $(x+y) = \sqrt{580 + 576} = \sqrt{1156} = 34$

4. $x - y = 5; x^2 + y^2 = 377; x^2 + y^2 - 2xy = 25$

   $\therefore -2xy = -352$

   $2xy = 352$

   $\therefore xy = 176$

   $\therefore (x+y)^2 = 377 + 352 = 729 \Rightarrow (x+y) = 27$

   $x^2 - y^2 = 27 \times 5 = 135$

   Check: 256 - 121 = 135 QED

5. $(x-2) + (x-1) + x + (x+1) = 4x - 2 = 1250$

   $\therefore 4x = 1252$

   $x = 313$

   $\therefore$ Largest is **314**

6. The pairs satisfying the requirement that the difference of the number and the number obtained by interchanging the digits is 36 are the following

51	62	73	84	95
-15	26	37	48	59
36	36	36	36	36

   Out of these only the pair 73, 37 are prime numbers. This is then the correct choice

7. Let xy be the original number $= 10x + y$;

   $yx = 10y + x = 4(10x + y) - 24 = 40x + 4y - 24$

   $\Rightarrow 6y = 39x - 24$;

   $y = x + 7 \Rightarrow 6y = 6x + 42$

   $\therefore 39x - 24 = 6x + 42$

   $\therefore 33x = 66$

   $\therefore x = 2; y = x + 7 = 9$

   Original number is **29**

8. Let $xy = 10x + y$ be the number. It is given that $(10x+y) - (10y+x) = 54$ and also $x = 3y$

   $\therefore 9x - 9y = 27y - 9y = 18y = 54$

   $\therefore y = 3, \quad x = 9$

   The no is 93

   Product required = 27

9. Let xyz the number. Middle no y =?

   a) $y = \dfrac{x+z}{3}$
   b) $x + y + z = 16$

   c) $zyx - xyz = 198$

   a) $3y = x + z$
   b) $(x+z) + y = 3y + y = 16$

   $\therefore y = 4; \quad x + z = 12$

   c) $100z + 10y + x - 100x - 10y - z = 99(z - x) = 198$

   $\left.\begin{array}{r}\therefore (z-x) = 2 \\ x + z = 12\end{array}\right\}\quad \therefore 2z = 14 \\ z = 7$

   $x = 5$

   The number is **547**

10. Let $\dfrac{a}{b}$ be the fraction. It is not mentioned whether (A) numerator > denominator or (B) num < den try choice (B) first

    (B) Fraction is $\dfrac{a}{a+7} = \dfrac{a}{b}$. When denominator is increased by 3 it becomes $\dfrac{a}{a+10} = \dfrac{c \cdot a}{d \cdot b}; \dfrac{c}{d} = \dfrac{9}{10}$

    $\Rightarrow \dfrac{a}{a+7} - \dfrac{a}{a+10} = \dfrac{9}{10} \Rightarrow \dfrac{a^2 + 10a - a^2 - 7a}{a^2 + 17a + 70} = \dfrac{9}{10}$

    $\Rightarrow 30a = 9a^2 + 153a + 630$

    $9a^2 + 123a + 630 = 0$

    $\Rightarrow 3a^2 + 41a + 210 = 0$

    This is not giving integral solutions

    So choice A must be correct Given

    (A) $\dfrac{a}{b} = \dfrac{b+7}{b}; \dfrac{c}{d} = \dfrac{b+7}{b+3}; \dfrac{b+7}{b} - \dfrac{b+7}{b+3} = \dfrac{9}{10}$

Word Problems on Numbers 55

$$\Rightarrow \frac{b^2 + 10b + 21 - b^2 - 7b}{b(b+3)} = \frac{9}{10}$$

$\Rightarrow 30b + 210 = 9b^2 + 27b$

$\Rightarrow 9b^2 - 3b - 210 = 0$

$\Rightarrow 3b^2 - b - 70 = 0$

$\Rightarrow 3b^2 - 15b + 14b - 70 = 0$

$\Rightarrow 3b(b-5) + 14(b-5) = 0$

Only b = 5 is admissible

Given ratio $= \dfrac{a}{b} = \dfrac{5+7}{b} = \dfrac{12}{5}$ ans

Check $\dfrac{12}{5} - \dfrac{12}{8} = 12 \cdot \dfrac{8-5}{40} = \dfrac{36}{40} = \dfrac{9}{10}$

```
 pr.
 3  × 210
 6  × 105
 9  × 70          Sum
15  × 42        ...... 57
18  × 35        ......
21  × 30        ...... 51
 -pr.
 3  × 70
 6  × 35
15  × 14   → diff = 1
```

11. Let $\dfrac{a}{b}$ be the fraction; given $\dfrac{a}{b} = \dfrac{3}{4}$ also $\dfrac{a-5}{b} = \dfrac{1}{6} \cdot \dfrac{a}{b}$

∴ $6a - 30 = a$

∴ $5a = 30$

∴ $a = 6$

Check $\dfrac{a}{b} = \dfrac{6}{8} = \dfrac{3}{4}$

$\dfrac{6-5}{8} = \dfrac{1}{8} = \dfrac{1}{6} \cdot \dfrac{1}{8}$

(For aliter solution see after 73, 72 aliter)

12. Let $\dfrac{a}{b}$ be the fraction;

Given $\dfrac{a-2}{b} = \dfrac{2}{3} \dfrac{a}{b} \rightarrow$ (1) & $\dfrac{a}{b-3} = \dfrac{1}{2} \rightarrow$ (2)

From (1) $3(a-2) = 2a \Rightarrow a = 6$;

From (2) $2a = b - 3 = 12$

∴ $b = 15$

∴ given fraction = 6/15

13. Let a, b, c be the numbers given
    $a^2 + b^2 + c^2 = 565 \rightarrow (1)$
    $ab + bc = ca = 558 \rightarrow (2)$
    Then $(a+b+c)^2 = a^2 + b^2 + c^2 + 2ab + 2bc + 2ca$
    $\qquad = 565 + 1116 = 1681$
    $\therefore (a + b + c) = \sqrt{1681} = 41$

14. Let xy be the number. Its value is $(10x+y)(yx)$ is the reversed number; It's value is $10y+x$

    Their difference is $10(x-y) + (y-x) = 9(x-y)$

    [In case we started with y x the difference is $10(y-x) + 9(x-y) = 9(y-x)$]

    Both are always divisible by 9

15. subtracting:

    $\begin{array}{cccccc} 10^5 & 10^4 & 1000 & 100 & 10 & 1 \text{ (units)} \\ f & e & d & c & b & a \\ a & b & c & d & e & f \end{array}$

    $10^5(f-a) + 10^4(e-b) + 10^3(d-c) + 100(c-d) + 10(b-e) + (a-f)$
    $= 99999(f-a) + 9990(e-b) + 900(d-c)$

    100000 10000
      -1   -10
    99999   9990

    1000 - 100 = 900

    Note:- Each multiple is divisible by 9 not 90 (first one) 99 (2nd one)

    $\therefore$ Difference is divisible by 9

16. $\begin{array}{ccccccc} 10^6 & 10^5 & 10^4 & 10^3 & 10^2 & 10 & 1 \\ g & f & e & d & c & b & a \\ -(a & b & c & d & e & f & g) \end{array}$

    $= 10^6(g-a) + 10^5(f-b) + 10^4(e-c) + 10^3(d-d) + (a-g) + 10^2(c-e) + 10(b-f)$

    $= 999999(g-a) + 99990(f-b) + 9900(e-c) + 1000(d-d)$

    Each of the terms is divisible by 99

    Hence the difference is divisible by 99

17. $p - 4 = q + 4 = 2r = \dfrac{S}{2} = $ say x

$\therefore p = x + 4; q = x - 4; r = \dfrac{x}{2}; s = 2x$

Then $p + q + r + s = (x-4) + (x+4) + \dfrac{x}{2} + 2x = 18$

$\Rightarrow 4x + \dfrac{x}{2} = \dfrac{9x}{2} = 18$

$\therefore x = 4$

The numbers are $p = 8, q = 0, r = 2s = 8$

Difference required = 8 - 0 = **8**

Key

1)	c	2)	c	3)	c	4)	b	5)	a	6)	c
7)	b	8)	c	9)	b	10)	d	11)	c	12)	b
13)	b	14)	d	15)	b	16)	c	17)	d		

7

Indices and Surds

Introduction

If 'x' is a rational number and 'n' a positive integer such that $x^{\frac{1}{n}} = \sqrt[n]{x}$ is rational then $\sqrt[n]{x}$ is called a surd of order 'n'

Ex.:1 $(625)^{0.13} \times 25^{0.24}$

$$= (25^2)^{0.13} \times 25^{0.24}$$
$$= 25^{0.26 + 0.24}$$
$$= 25^{0.5}$$
$$= 25^{\frac{1}{2}}$$
$$= 5$$

Ex.:2 Find x when $\left[27^{\left(\frac{x-1}{3}\right)} + 27^{\left(\frac{x+1}{3}\right)}\right] = 2430$

$$\text{LHS} \Rightarrow 27^{\frac{x}{3}} \left(27^{-\frac{1}{3}} + 27^{\frac{1}{3}}\right)$$
$$= 3^x \left(\frac{1}{3} + 3\right)$$
$$= 3^x \frac{10}{3} = \text{RHS}$$
$$= 2430$$
$$\therefore 3^x = 243 \times 3$$
$$3^x = 3^6$$
$$\therefore x = 6$$

Ex.:3 Find the value of $\left(3^{\frac{1}{4}} - a\right)\left(3^{\frac{3}{4}} + 3^{\frac{1}{2}}a + 3^{\frac{1}{4}}a^2 + a^3\right)$

When $a = 2^{\frac{1}{4}}$.

Sol.: Put $x = 3^{\frac{1}{4}}$:

$$\begin{aligned}
\text{Given} &= (x-a)(x^3 + x^2 a + xa^2 + a^3) \\
&= (x-a)\left[x^2(a+x) + a^2(x+a)\right] \\
&= (x-a)(x+a)(x^2 + a^2) \\
&= (x^2 - a^2)(x^2 + a^2) \\
&= x^4 - a^4 \\
&= 3^{\frac{1}{4} \cdot 4} - 2^{\frac{1}{4} \cdot 4} \\
&= 3 - 2 \\
&= 1
\end{aligned}$$

Ex.:4 Find which is larger, $\sqrt{7}$ or $\sqrt[3]{11}$?

Sol.: To compare we have to convert into **surds of the same order:** LCM of 2 & 3 is 6. So we have to **convert into surds of order 6**

$$\begin{aligned}
\text{Now} \quad \sqrt[2]{7} &= 7^{\frac{1}{2} \times \frac{3}{3}} \\
&= 7^{\frac{3}{6}} \\
&= \sqrt[6]{7^3} = \sqrt[6]{343} \quad \& \\
\sqrt[3]{11} &= 11^{\frac{1}{3}} = 11^{\frac{1}{3} \times \frac{2}{2}} \\
&= 11^{\frac{2}{6}} \\
&= \sqrt[6]{121}
\end{aligned}$$

Comparing $\sqrt[6]{343}$, $\sqrt[6]{121}$, obviously $\sqrt[6]{343}$ is larger

Thus $\sqrt{7} > \sqrt[3]{11}$

Ex.:5 If m & n are whole numbers such that $m^n = 169$, find $(n-3)^{n+3}$

Sol.: obviously $m^n = 13^2$

$$\begin{aligned}
\text{i.e} \quad m &= 13 \ \& \ n = 2 \\
\text{Then} \quad (m-3) &= 10 \ \& \ (n+3) = 5 \\
\therefore (m-3)^{n+3} &= 10^5 \\
&= 100000
\end{aligned}$$

Ex.:6 Simplify $\left(\dfrac{x^b}{x^c}\right)^{b+c-2a} \cdot \left(\dfrac{x^c}{x^a}\right)^{c+a-2b} \left(\dfrac{x^a}{x^b}\right)^{a+b-2c}$

Sol.: $x^{(b^2+bc-2ab)} - (cb + c^2 - 2ac) + (c^2 + ac - 2bc)$
$- (ac + a^2 - 2ab) + a^2 + ab - 2ac - (ba + b^2 - 2bc)$

$$= x^0$$
$$= 1$$

Ex.: 7 If $3^x = 5^y = 15^{-z}$. What is $\left(\dfrac{1}{x} + \dfrac{1}{y} + \dfrac{1}{z}\right) = ?$

Sol.: Let each be $= a$ Then $3^x = a$

$$\Rightarrow 3 = a^{\frac{1}{x}}$$
$$5^y = a$$
$$\therefore 5 = a^{\frac{1}{y}} \ \& \ 15^{-z} = a$$
$$\therefore 15 = a^{-\frac{1}{z}}$$

Now $3 \times 5 = 15$

$$\therefore a^{\frac{1}{x}} \cdot a^{\frac{1}{y}} = a^{-\frac{1}{z}} \text{ or } \frac{1}{x} + \frac{1}{y} = -\frac{1}{z} \text{ or}$$

$$\left(\frac{1}{x} + \frac{1}{y} + \frac{1}{z}\right) = 0$$

In Unit 6 we have already seen various problems on powers.

x^a x: base, 'a' is 'power' (x is raised to the ath power. If x is a rational number, and 'a' a positive integer and if $x^{\frac{1}{a}}$ is irrational; $x^{\frac{1}{a}}$ is called a surd of order 'a' ex ... $3^{\frac{1}{3}}$ is a surd; $(3 + \sqrt{7})^{\frac{1}{4}}$ also a surd etc

Exercises

1. If p and q are whole numbers such that $p^q = 169$. What is the value of $(p-1)^{q+1}$?

    (a) 216 (b) 432 (c) 864 (d) 1728

2. If x and y are whole numbers such that xy = 169, find the value of $\dfrac{1}{(x-4)^4}$

(a) - 13 (b) - 81 (c) 729 (d) 6561

3. If x and y are whole numbers xy = 529; find $\left(\dfrac{2^4 x}{y-7}\right)$?

(a) 17 (b) 19 (c) 23 (d) 27

4. What is the value of $\dfrac{625^x (50000)^{x-2}}{(125)^{2x} \, 16^{x-2}}$ when x = 4?

(a) 5 (b) 16 (c) 25 (d) 125

5. x and y are real numbers what are their values if $6^{1.8} \times 12^{3.2} \div 54 \times 24^2 = 3^x 2^y$?

(a) (3, 9.1) (b) (4, 13.2) (c) (5, 16.4) (d) (7, 18.8)

6. If $49\sqrt{7} \times 7^4 \div 7^{-5/2} = 7^{x+3}$, What is x?

(a) 2 (b) 4 (c) 6 (d) 8

7. $7^{2x+3} - 7^{2x+1} = 2^4 \times 3$. What is x?

(a) $\dfrac{-1}{3}$ (b) $\dfrac{-1}{2}$ (c) 0 (d) $\dfrac{1}{3}$

8. Given $7^x - 7^{x-2} = 2352$, find x^x =?

(a) 64 (b) 128 (c) 256 (d) 512

9. One of two surds whose Arithmetic mean is $4 + 3\sqrt{2}$ is $3 + 2\sqrt{2}$. Find the other surd

(a) $5 - \sqrt{2}$ (b) $5 + \sqrt{2}$ (c) $5 + 3\sqrt{2}$ (d) $5 + 4\sqrt{2}$

10. If $\sqrt{20 + x\sqrt{11}} = 3 + \sqrt{11}$. Find x

(a) 2 (b) 4 (c) 6 (d) 8

Solutions

1. $p^q = 169$ ∴ p = 13; q = 2 ∴ (p - 1) = 12; and q + 1 = 3

∴ $(p-1)^{q+1} = 12^3 = 1728$

2. 169 is a number with factor 13; $13^2 = 169$

$$\therefore x = 13; \quad y = 2$$

$$\therefore \frac{1}{(x-4)^{-4}} = \frac{1}{(13-4)^{-4}} = \frac{1}{9^{-4}} = \frac{1}{(81 \times 81)^{-1}} = 81 \times 81$$

$$= 6561$$

3. $x^y = 529$ is a perfect square or a prime number product etc

   To find whether 529 is a perfect square so $23^2 = 529$

   $$\therefore x = 23 \quad y = 2$$

   ```
   5 29|23
        4
   43 |129
      |129
      | x
   ```

   Then $\dfrac{2^{4x}}{(y-7)} = \dfrac{16 \times 23}{23 - 7} = \dfrac{16 \times 23}{16} = 23$

4. $\dfrac{(5^4)^x \, 5^{x-2} \, (10^4)^{x-2}}{(5^3)^{2x} \, (2^4)^{x-2}} = \dfrac{5^{4x+x-2-6x} \, 5^{4x-8} \, 2^{4x-8}}{2^{4x-8}} = 5^{3x-10}$

   When $x = 4$

   This is $5^{12-10} = 5^2 = 25$

5. $(3^{1.8} \times 2^{1.8} \times 3^{3.2} \times 2^{6.4}) \div (2 \times 3^3) \times (2^3 \times 3)^2$

   $= $ Given $3^x \, 2^y$

   $\Rightarrow (3^5 \, 2^{8.2}) \div (2 \times 3^3) \times (2^3 \times 3)^2 = 3^x \, 2^y$ using BODMAS rule

   L.H.S $(3^{5-3} \, 2^{8.2-1}) \times 2^6 \times 3^2 = 3^2 \, 2^{7.2} \times 2^6 \times 3^2$

   $= 3^4 \times 2^{13.2} = 3^x \, 2^y$

   $\therefore x = 4$ and $y = 132$

6. LHS: $7^2\sqrt{7} \times (7^4 \div 7^{-5/2})$

   $= 7^{2.5} \times 7^{4+5/2}$

   $= 7^{2.5} \times 7^{13/2}$

   $= 7^{(2.5+6.5)} = 7^9$

   R.H.S $= 7^{x+3}$ ∴ $x + 3 = 9$ ∴ $x = 6$

7. Given L.H.S $= 7^{2n+1}(7^2 - 1)$

   $= 7^{2n+1} \times 48$

   R.H.S $= 48$ ∴ $7^{2n+1} = 1 = 7^0$

   Equaling indices $2n + 1 = 0$

   ∴ $n = -1/2$

   Check L.H.S $7^2 - 7^0 = 49 - 1 = 48$

   R.H.S $= 48$

8. $7^{n-2}[7^2 - 1] = 2352$

   $7^{n-2} = \dfrac{2352}{48} = \dfrac{588}{12} = \dfrac{147}{3} = 49 = 7^2$

   ∴ $(n - 2) = 2$ ∴ $n = 4$ ∴ $n^n = 4^4 = 256$

9. If x is the other surd; then $\dfrac{3 + 2\sqrt{2} + x}{2}$

   $4 + 3\sqrt{2}$

   ∴ $3 + 2\sqrt{2} + x = 8 + 6\sqrt{2}$

   ∴ $x = 5 + 4\sqrt{2}$

10. Squaring $20 + x\sqrt{11} = 9 + 11 = 6\sqrt{11}$

    ∴ $x = 6$

Key

1)	d	2)	d	3)	c	4)	c	5)	b	6)	c
7)	b	8)	c	9)	d	10)	c				

8

Logarithms

Exercises

Note: When base is not mentioned it is taken as 10

1. If $\log_{10} 2 = 0.30103$, find $\log_{10} 4$

    (a) 0.60206 (b) 1.30103 (c) 2.30103 (d) 2.60206

2. If $\log_{10} 27 \simeq 1.4310$ and $100^x = 27$; find x

    (a) 0.7155 (b) 1.4310 (c) 2.1465 (d) 2.4310

3. If $\log_{10} 2 = 0.30103$, find the number of digits in 2^{58}?

    (a) 18 (b) 23 (c) 25 (d) 27

4. Find x if $\log 4 + \log(6x - 20) = 2\log(x - 3) + 1$

    (a) 5 or $\dfrac{17}{5}$ (b) 5 or $\dfrac{27}{5}$ (c) 10 or $\dfrac{27}{5}$ (d) 10 or $\dfrac{27}{5}$

5. Find $\log_{625} 5$

    (a) $\dfrac{-1}{4}$ (b) $\dfrac{-1}{3}$ (c) $\dfrac{1}{4}$ (d) 0.3

6. Find $\log_6 \dfrac{1}{216}$

    (a) -3 (b) 0.3 (c) $\dfrac{1}{3}$ (d) 3.3

7. Find $\log_3 0.00243$ if $\log_3 = 0.4771$

   (a) -5.4800 (b) 0.4800 (c) 5.4800 (d) 10.4800

8. Find $\frac{1}{5} \log_{10} 1024 + \log_{10} 2.5$

   (a) -1 (b) 0 (c) 1 (d) 2

9. If $\log 625 + \log 16 = x$, find x

   (a) 1 (b) 2 (c) 3 (d) 4

10. Find x given $\log_7 (x^3 - x^2) - \log_7 (x - 1) = 2$

    (a) 2 (b) 4 (c) 6 (d) 8

11. $\left[\dfrac{1}{\log_x(yz) + 1} + \dfrac{1}{\log_y(zx) + 1} + \dfrac{1}{\log_z(xy) + 1} \right] = ?$

    (a) -1 (b) 0 (c) 1 (d) 2

Solutions

1. $\log_{10} 2 = 0.30103$; $\log_{10} 4$

   $\log_{10} 2^2 = 2 \log_{10} 2 = 0.60206$

2. $\log_{10} 27 = 1.4310$ & $100^x = 27$

   $10^{1.4310} (10^2)^x$

   $\therefore 2x = 1.4310$

   $\Rightarrow x = 0.7155$

3. Let $2^{58} = x$; $\log_{10} 2 = \left(\dfrac{\log_{10} x}{58} \right)$

   $\Rightarrow 58 \log_{10} 2 = \log_{10} x$

   $\Rightarrow \log_{10} x = 58 \times 0.30103 = 17.4597$

   Hence number of digits in 2^{58} will be $17 + 1 = 18$

4. $\log 4 + \log(6x - 20) = 2\log(x - 3) + 1 = x\log(x-3)^2 + \log 10$

$\Rightarrow \log 4(6x - 20) = \log(x-3)^2 + \log 10 \Rightarrow \log 10(x-3)^2$

$\Rightarrow 4 \cdot (6x - 20) = (x - 3)^2 \cdot 10$

$\Rightarrow 2(6x - 20) = 5(x - 3)^2$

$12x - 40 = 5x^2 - 30x + 45$

$5x^2 - 42x + 85 = 0$

$5x^2 - 25x - 17x + 85 = 0$

$5x(x - 5) - 17(x - 5) = 0$

$(5x - 17)(x - 5) = 0$

$5x = 17$ or $x = 5$

$x = 17/5$ or 5

Check:

L.H.S $2\left(6 \times \dfrac{17}{5} - 20\right) = 2\left(\dfrac{102}{5} - 20\right) = 2 \times \dfrac{2}{5} = \dfrac{4}{5}$

$\left(x = \dfrac{17}{5}\right)$: R.H.S $5\left[\dfrac{17}{5} - 3\right]^2 = 5 \times \dfrac{2}{5} \times \dfrac{2}{5} = \dfrac{4}{5}$

$(x = 5)$: R.H.S $2(30 - 20) = 20$ R.H.S $5(5 - 3)^2 = 20$

5. Given $\Rightarrow 625^x = 5 \Rightarrow 625^x = (5^4)^{1/4}$

$= (625)^{1/4}$ $\therefore x = \dfrac{1}{4}$

6. Let $\log_6\left(\dfrac{1}{216}\right) = x$ then

$6^x = \dfrac{1}{216} = \dfrac{1}{(6)^3} = 6^{-3} \Rightarrow x = -3$

7. Let $\log_3(0.00243)$ be x. then $3^x = 243 \times 10^{-5} = 3^5 \times 10^{-5}$

Taking logs on both sides to the base 10

$x \log_3 = \log \dfrac{3^5}{10^5} = 5\log_3 - 5\log 10$

Logarithms 67

$$\therefore x = \frac{5\log 3 - 5}{\log 3} = 5 - \frac{5}{\log 3} \Rightarrow 5 - \frac{5}{0.4771}$$

$$= 5 - 10.4799 = -5.4799 = \bar{5}.4799 \simeq \bar{5}.4800$$

8. $\Rightarrow \dfrac{1}{5}\log_{10} 4^5 + \log_{10} 2.5 = \dfrac{1}{5} 5\log_{10} 4 + \log_{10} 2.5$

$\Rightarrow \log_{10} 4 + \log_{10} 2.5 = \log_{10} 10 = 1$

9. Let $\log_{10} 625 + \log_{10} 16 = \log_{10} 10000 = \log_{10} 10^4 = 4$

10. Given $= \log_7 x^2 + \log_7 (x-1) - \log_7 (x-1) = 2$

    $\therefore x^2 = 7^2 \quad \therefore x = 7$

    Check $\log_7 (343 - 49) - \log_7 6 = \log_7 294 - \log_7 6$

    $= \log_7 \dfrac{294}{6} = \log_7 49 = 2$ QED

11. Given $= \sum \dfrac{1}{\log_x zy + \log_x x} = \sum \dfrac{1}{\log_x xyz}$

    $\Rightarrow \log_{xyz} x + \log_{xyz} y + \log_{xyz} z = \log_{xyz} xyz = 1$

Key

1)	a	2)	a	3)	a	4)	c	5)	c	6)	a	7)	b
8)	c	9)	d	10)	a	11)	c						

9

Ratio and Proportion

Exercises

1. Two numbers are in the ratio of 2:3. When 6 is subtracted from each their ratio becomes 12:19. What is the larger number?

   (a) 36   (b) 42   (c) 57   (d) 63

2. One Rupee coins, fifty paise coins and twenty five paise coins in the ratio of 3:4:5 are collected in a piggy Bank. If the total amount is Rs. 500/-. How many half rupee coins are there?

   (a) 160   (b) 240   (c) 320   (d) 400

3. Salaries of A and B which are in the ratio of 3:4 are hiked by Rs. 3000/- each on first October. What is 'B's salary after the hike. If the new ratio is 7:9?

   (a) 18000   (b) 21000   (c) 24000   (d) 27000

4. Rs. 570/- is to be distributed to three persons such that "p" gets 3/4 of what 'Q' gets and "Q" gets 1/3 of what "R" gets. What is the share of "P"?

   (a) 90   (b) 120   (c) 180   (d) 360

5. The sum of three numbers (a, b, c) is 126, a : b is 3 : 4 and b: c = 5 : 7. What is "B"?

   (a) 30   (b) 40   (c) 56   (d) 63

6. Two numbers are 10% and 15% less than a third number. The ratio of the two numbers is

   (a) 9:16   (b) 9:14   (c) 13:16   (d) 18:17

7. The number of seats in a college are in the ratio of 3:4:5. In mathematics, physics & Biology. If the seats are to be increased by 20%, 25%, 40%. What is the new ratio?

   (a) 9:12:13   (b) 9:10:14   (c) 18:20:30   (d) 18:25:35

8. If in a 50l admixture the ratio of milk to water is 3:2. How much water is to be added to make the ratio 1:1?

   (a) 5l   (b) 10l   (c) 15l   (d) 20l

9. 30% of 'a' is 70% 'b'. What is a:b?

   (a) 3:7   (b) 6:7   (c) 7:3   (d) 7:6

10. If x:y is 6:7 and y is to z is 8:9. What is x: y: z?

    (a) 42: 54: 59   (c) 48: 54: 60
    (b) 42: 52: 62   (d) 48: 56: 63

11. Two numbers bear a ratio of 2:3; their LCM is 36. What are the numbers?

    (a) 12, 18   (b) 12, 24   (c) 18, 24   (d) 18, 36

12. An alloy contains constituents c_1, c_2 in the ratio of $c_1 = c_2 = 8:3$ (to be melted). The c_2 required with 40 kgs of c_1 is

    (a) 5kg   (b) 10kg   (c) 15kg   (d) 20kg

13. Constituent c_1 and constituent c_2 are 20 and 10 times heavier than water. What is the ratio in while they must be mixed to get an alloy 15 times heavier than water?

    (a) 1:1   (b) 1:2   (c) 1:3   (d) 3:1

14. If is the CSE class boys and girls are in the ratio of 5:3 and the strength of girls is 24, what is the class strength?

    (a) 57   (b) 60   (c) 64   (d) 72

15. In a village 8% of the Hindus were equal in no to 48% of the Muslims. What is the ratio of Hindus to Muslims?

    (a) 4:1   (b) 6:1   (c) 8:1   (d) 16:1

16. A vehicle covers 120 kms in 4 hours what is the velocity of the vehicle? If the speed is doubled what will be the time to cover the same distance?

    (a) Time is doubled   (c) Time is 3/4 the of earlier time
    (b) Time is same   (d) Time is halved

Solutions

1. Let $\dfrac{2x}{3x}$ be the ratio of the numbers $\left(=\dfrac{2}{3}\right)$;

   $$\dfrac{2x-6}{3x-6} = \dfrac{12}{19}$$

   ∴ 38x - 114 = 36x - 72

   2x = 42   ∴ x = 21

   ∴ Large number is 3x = 63 ans

2. Let 3x, 4x, 5x be the number of 1 rupee, 50ps and 25ps coins then

   $(3x)(1) + 4x\left(\dfrac{1}{2}\right) + 5x\left(\dfrac{1}{4}\right)$ = Rs. 500/- or

   $3x + 2x + \dfrac{5}{4}x = \dfrac{12x + 8x + 5x}{4} = \dfrac{25x}{4}$ = Rs. 500/-

   Then x = $\dfrac{500 \times 4}{25}$ = 80;

   Number of $\dfrac{1}{2}$ rupee coins = 4x = 320

   Check 240 + 160 + 100 = Rs. 500/-

3. Let 3x, 4x be the original salaries

   3x + 3000 : 4x + 3000 = 7 : 9

   $\therefore$ 27x + 27000 = 28x + 21000

   $\therefore$ x = 6000

   $\therefore$ new salary, after hike of B: 4x + 3000

   = 24000 + 3000

   = Rs. 27000/- p.m

4. Let R get x; Q gets 1/3 x; and P gets $\dfrac{3}{4} \times \dfrac{1}{3} x = \dfrac{x}{4}$

   $$\text{Total} = \dfrac{x}{4} + \dfrac{x}{3} + x = \dfrac{3 + 4 + 12}{12} x = \dfrac{19x}{12} = 570$$

   $$\therefore x = \dfrac{12}{19} \times 570 = 360$$

   Hence P gets $\dfrac{x}{4} = \dfrac{360}{4} = 90$

5. a : b = 3 : 4 $\Rightarrow$ 15 : 20

   b : c = 5 : 7

   $\therefore$ b : c = 20 : 28

   $\therefore b = \left(\dfrac{20}{15 + 20 + 28}\right) \times 126$

   $= \dfrac{20}{63} \times 126 = 40$

6. Let x be the third number. Then the numbers are $\left(x - \dfrac{10}{100}x\right)$ and $\left(x - \dfrac{15x}{100}\right)$

   i.e., $\dfrac{90x}{100}$ and $\dfrac{85x}{100}$

   Their ratio is $\dfrac{90}{85} = \dfrac{18}{17}$

7. Let 3x, 4x, 5x be the seats, increased seats are $\left(3x+\frac{1}{5}3x\right)$, $\left(4x+\frac{1}{4}.4x\right)$ and $\left(5x+\frac{2}{5}4x\right)$ or $\frac{18x}{5}$: 5x : 7x

The new ratio will be = $\frac{18x : 25x : 35x}{5}$ or 18: 25: 35

8. Let 3x, 2x be the values in 50l of admixture then milk is 30l and water is 20l in the 50l of admixture to make the ratio 1:1, 10l of water is to be added 30l: 30l or 1:1 in the 60l of admixture

9. $\frac{3}{10}a = \frac{7}{10}b$

$\therefore \frac{a}{b} = \frac{7}{10} \times \frac{10}{3} = \frac{7}{3}$

10. x: y = 6:7 or 48 : 56

  y : z = 8 : 9 = 56 : 63
  $\therefore$ x : y : z = 48 : 56 : 63

11. Let the numbers be 2x, 3x

2x : 3x. Their LCM? x $\underline{|\ 2x, 3x}$
                         2, 3

$\therefore$ LCM is 6x = 36 (given)

$\therefore$ x = 6

$\therefore$ Numbers are 12, 18

12. $c_1 : c_2 = 8 : 3 = 40 : x$

$\therefore x = \frac{3 \times 40}{8}$ = **15 kg**

13. $c_1 = 20w$, $c_2 = 10w$ if 1 gm of c_1 is mixed with a gram of c_2 to get (a + 1) gram of alloy; then (20 w) (1) + (10 w) (a) = (a + 1) 15 w

$\Rightarrow$ 20 – 15 = 15a – 10a = 5a
$\therefore$ a = 1

Thus c_1, c_2 should be mixed in the ratio of 1: 1

Check: $(20w) 1 + (10w) 1 = (1 + 1) (15w)$

Or $20w + 10w = 30w$

14. $\dfrac{8}{3} \times 24 = 64$

15. Let h and m be the numbers

$\dfrac{8}{100} h = \dfrac{48}{100} m \qquad \therefore \dfrac{h}{m} = \dfrac{48}{8} = 6:1$

16. Distance = 120 km if speed is 30 kh/hr → time is 4 hours, if the speed is doubled = 60 kh/hr, time is 2 hours (halved) i.e if speed increases time to cover the same distance decreases $x \,\alpha\, \dfrac{1}{y}$ (inversely proportional)

$\boxed{\text{Key}}$

1)	d	2)	c	3)	d	4)	a	5)	b	6)	d
7)	d	8)	b	9)	c	10)	d	11)	a	12)	c
13)	a	14)	c	15)	b	16)	d				

10

Percentages

Exercises

1. 0.50 ps is the charge for a page in a Xerox copy shop. 4% discount is allowed after 500 pages. What is the cost to copy 4500 pages?

   (a) Rs. 2030/-   (b) Rs. 2070/-   (c) Rs. 2170/-   (d) Rs. 2250/-

2. A house wife saved Rs. 3/- in an item on sale. She paid Rs. 33/- for the item. What is the percent saving in the transaction?

   (a) 8%   (b) 8.33%   (c) 9%   (d) 9.5%

3. Anil purchases items worth Rs. 7250/- He gets a rebate of 8% on it. After getting the rebate, he pays sales 12%. How much amount has he to pay finally?

   (a) Rs. 6500/-   (c) Rs. 7250/-
   (b) Rs. 6670/-   (d) Rs. 7470.40/-

4. Which one of the following shows the largest percentage?

   (a) $\dfrac{410}{550}$ (b) $\dfrac{440}{540}$ (c) $\dfrac{550}{700}$ (d) $\dfrac{490}{650}$

5. If the difference between a number and its $\dfrac{3}{8}$ th is 860 is 25% of that number?

   (a) 86   (b) 172   (c) 344   (d) 688

6. What is the percentage of numbers from 1 to 80 having squares that end with the digit 9?

   (a) 5%   (b) 10%   (c) 20%   (d) 40%

7. The product of two numbers is 13500. 4.5% of one number is 7.5% of other. What are they?

   (a) 90, 150   (b) 150, 90   (c) 180, 300   (d) 300, 180

8. The votes polled in an election are 5672, 4341 and 22, 287 by A, B, and C. What is the percentage vote of the winning candidate?

   (a) 19%   (b) 37%   (c) 59%   (d) 69%

9. Population increased from 234000 to 385000 in a decade. What is the percentage increase per year?

   (a) 5%   (b) 6.453%   (c) 7.5%   (d) 8%

10. A student multiplied a number by $\frac{3}{5}$ instead of by $\frac{5}{3}$. What is the percentage error in the calculation?

    (a) 34%   (b) 44%   (c) 54%   (d) 64%

11. A person purchased items worth Rs. 40/- out of which 60 paid went on sales, tax on taxable items. The tax rate was 6%. What was the cost of tax free items?

    (a) Rs. 25.6/-   (b) Rs. 27.00/-   (c) Rs. 29.40/-   (d) Rs. 30/-

12. A bats man secured 70 runs out of which one was a boundary and six were sixers. What percent of his total score did he make by running between the wickets?

    (a) 35%   (b) 37.9%   (c) 42.87%   (d) 49%

13. After selling 70% of apples a fruit seller still has 45 apples left with him. What was the original number of apples?

    (a) 75   (b) 100   (c) 150   (d) 175

14. A person spends 60% of his income. His saving is Rs. 1200/-. What is the expenditure?

    (a) 1200 (b) 1500 (c) 1800 (d) 2000

15. 5% of a village population died with plague 20% of the remaining people left the village. The population is then only 6080, what is the original number of inhabitants on the village?

    (a) 4000 (b) 6080 (c) 8000 (d) 16000

16. A sales man gets 6% commission up to Rs 10,000 and 5% an the sales exceeding this. After deducting his commission he remits Rs. 37,900/- the company. What are the total sales in that year?

    (a) 15000 (b) 24000 (c) 38000 (d) 40000

17. The price of kerosene is increased by 20%. What should be the decrease in the consumption by a person so that his expenditure does not increase?

    (a) 8% (b) 8.3% (c) 16.66% (d) 20%

18. How many litres of pure acid are to be added to a 10 litre 5% solution of acid and water, to make it a 10% solution?

    (a) 0.5l (b) 0.55l (c) 0.6l (d) 0.66l

| Solutions |

1. First 500 pages → $\frac{1}{2} \times 500$ = Rs. 250/-

   Discount for 100ps → 4 for 50ps ⇒ 2ps 48 is the rate for the other pages

   $= \frac{48}{100} \times 4000$ = Rs. 1920/-

   Total cost = 1920 + 250 = Rs. 2170/-

2. If x is the original price she purchased for (x − 3) = 33

   ∴ x = 36

   ∴ Saving = $\frac{3}{36} \times 100 = \frac{100}{12} = \frac{25}{3}$ = 8.33%

Percentages

3. Rs. 7250/-; Rebate = $\dfrac{8}{100} \times 7250$ = Rs. 580/-

   $\therefore$ (7250 - 580) = Rs. 6670/- is the amount on which he has to pay a sales tax of 12%

   $\therefore \dfrac{112}{100} \times 6670 = 7470.40$ is the amount he has to pay

4. These are 74.54, 81.48, 78.57 and 75.38

5. $\left(x - \tfrac{3}{8}x\right) = \dfrac{5}{8}x = 860 \quad \therefore x = \dfrac{860 \times 8}{5}$

   $\therefore \dfrac{25}{100}x = \dfrac{25}{100} \times \dfrac{860 \times 8}{5}$

   $= \dfrac{40 \times 860}{100} = 344$

6. Listing out 1 to 10: 3 and 7; for 11 to 20 = 13 and 17 and so on, so upto 80 = 16 out of 80

   $\therefore$ Percentage = $\dfrac{16}{80} \times 100 = 20\%$

7. xy = 13500; $\dfrac{4.5}{100}x = \dfrac{7.5}{100}y$

   $\therefore \dfrac{x}{y} = \dfrac{75}{45} = \dfrac{5}{3}$

   $\therefore x = \dfrac{5}{3}y$

   $\therefore \dfrac{5}{3}y \cdot y = 13500$

   $\therefore y^2 = 13500 \times \dfrac{3}{5} = 8100$

   $\therefore y = 90 \qquad x = 150$

8.
   ```
   5672
   4341
   22287
   -----
   32300
   ```
   $\therefore \dfrac{22287}{32300}\ 0.69 = 69\%$

9. $\begin{array}{r} 385000 \\ -234000 \\ \hline 151000 \end{array}$ $\quad \therefore \dfrac{151000 \times 100}{10 \times 234000} = \dfrac{1510}{234} = 6.453\%$

10. $\dfrac{\dfrac{5}{3} - \dfrac{3}{5}}{\dfrac{5}{3}} \times 100 \quad \dfrac{\dfrac{25-9}{15}}{\dfrac{5}{3}} \times 100 = \dfrac{16}{15} \times \dfrac{3}{5} \times 100$

$$= \dfrac{16}{25} \times 100 = 64\%$$

11. Let the amount of taxable items be Rs. x. Tax is 6% so tax is $\dfrac{6x}{100}$ is Rs. $\dfrac{60}{100}$

    $\therefore$ x = Rs. 10/-

    Worth of tax free items = 40 - 10 = Rs.30/-

    Price paid = 30 - 0.60 = Rs. 29.40/-

12. One boundary = 4 runs+ 6 sixes = 36.

    Actual runs = 70 - 40 = 30

    percentage = $\dfrac{30}{70} \times 100 = \dfrac{300}{7} = 42.87\%$

13. $\dfrac{30x}{100} = 45; x = \dfrac{100 \times 45}{30} = \dfrac{100 \times 3}{2} = 150\%$

14. Let income = x, Savings, 40x = 1200

    $\therefore$ Expend. $60x = \dfrac{1200 \times 60x}{40 x}; x = 1800$

15. 100% → 5% died 95% remaining; 80% (or) $\dfrac{80}{100} \times 95$ [20% of the remaining left i.e 19 out of total 100 left], 76% are remaining in the village this is equal to 6080

$$\therefore 100 \times \dfrac{6080}{76} = 8000$$

Check Total = 8000
Died = 400
Left = 20% of remaining or 7600 i.e 1520
Remaining = 6080

16. If Rs. x/- is the sales,

$$x - \left[\frac{6}{100} \times 10,000\right] + \left[\frac{5}{100}(x - 10,000)\right] = 37,900$$

$$\therefore x - \frac{60,000}{100} - \frac{5x}{100} + 500 = x - \frac{5x}{100} - 100 = 37,900$$

$$\therefore \frac{95x}{100} = 38,000$$

$$\therefore x = \left\{\frac{2000 \times 3,8000 \times 100}{95}\right\} = 40,000$$

17. $p \rightarrow \left(p + \frac{20p}{100}\right)$. Let c be the consumption and x% the reduction in consumption

$$\therefore P\left(1 + \frac{1}{5}\right)\left(c - \frac{xc}{100}\right)$$

Should be pc; or $pc\left(1 + \frac{1}{5}\right)\left(1 - \frac{x}{100}\right) = pc$

or $\frac{6}{5}\left(1 - \frac{x}{100}\right) = 1$ or $6 - \frac{6x}{100} = 5$

$$\therefore 1 = \frac{6x}{100} \quad \therefore x = \frac{100}{6} = 16.66\%$$

18. 10 l of 5% solution contains $\frac{5}{100} \times 10 = \frac{1}{2}l$ acid; if x l of pure acid is added then the acid content will be $\left(\frac{1}{2} + x\right)l$

$\therefore$ Total volume now is $(10 + x)l$

$\therefore$ New concentration = $\dfrac{\frac{1}{2} + x}{10 + x}$ given as 10%

$$= \frac{10}{100} = \frac{1}{10}$$

$$\therefore \left(\frac{\frac{1}{2} + x}{10 + x}\right) = \left(\frac{1}{10}\right)$$

$\therefore 5 + 10x = 10 + x$

$\therefore 9x = 5$

$\therefore x = \frac{5}{9} = 0.551$

$$\boxed{\text{Key}}$$

1)	c	2)	b	3)	d	4)	b	5)	c	6)	d
7)	b	8)	d	9)	b	10)	d	11)	c	12)	c
13)	c	14)	c	15)	c	16)	d	17)	c	18)	b

11

Averages

Exercises

1. The average of all the two digit numbers, which remain the same when the digits are interchanged is

    (a) 33   (b) 44   (c) 55   (d) 66

2. The average of a positive number and its square is equal to 3 times the numbers plus 3. What is the number?

    (a) 4   (b) 6   (c) 8   (d) 12

3. The average of 9 consecutive numbers is 24. What is the 4th no?

    (a) 17   (b) 19   (c) 21   (d) 23

4. The average of 5 consecutive odd numbers is 31. What is the average of the even numbers lying between the smallest and the largest odd numbers

    (a) 28   (b) 29   (c) 31   (d) 35

5. A family consists of grand parents, parents and children. Average age of grand parents is 69y, parents 34y, grand children 4 years. What is the average age of the family?

    (a) 32y   (b) $33\frac{3}{4}$ (c) $34\frac{1}{3}$ (d) $35\frac{2}{3}$

6. A library has an average of 48 visitors in an Sundays and 24 on working days. What is the average no of visitors in a month of 30 days beginning with a Sunday?

(a) 24 (b) 28 (c) 30 (d) 45

7. Batches of 60, 60 and 62 got average marks 49%, 45.5% and 42% receptively. What is the average marks of the I.T. class with these 3 sections?

   (a) 42.75% (b) 43.82% (c) 45.46% (d) 48.12%

8. A company produces 3000 items on an average in the 1st quarter of an year. If the yearly average should be 3750 items, what is the average it should produce in the remaining 3 Quarters?

   (a) 3850 (b) 3900 (c) 4000 (d) 4200

9. In 8 overs the average run rate was 3.5. What should be the run rate in the remaining 32 over to reach the target of 188 runs?

   (a) 4.2 (b) 4.8 (c) 5 (d) 5.5

10. Average of 5 numbers is 24, first number is 20% of the sum of the other four. What is the first number?

    (a) $16\frac{2}{3}$ (b) 17.5 (c) 18 (d) 20

11. The average weight of 10 persons increases by 1.5 kg, when a new person replaces another weighing 62 kg. What is weight of new person?

    (a) 69 kg (b) 73 kg (c) 77 kg (d) 80 kg

12. A cricketer has a certain average for 5 innings. In the 6th he secured 102 runs, average increases by 4 runs, what is new average?

    (a) 60 runs (b) 75 runs (c) 78 runs (d) 82 runs

13. A cricketer whose bowling average is 10.8 runs per wicket takes 3 wickets for 22 runs and there by decreases his average by 0.8. What is the no. of wickets taken by him till the last match?

    (a) 5 (b) 10 (c) 20 (d) 25

14. The average weight of P, Q, R is 81 kg. S joins and the average becomes 75 kg. Another person T who is '12' kg more than s replaces P and the average of Q, R, S, T is 74 kg. What is the weight of P?

(a) 57 kg   (b) 69 kg   (c) 73 kg   (d) 83 kg

15. The average marks of a class is 70. The average marks of boys is 49 and girls is 77. What is the ratio of boys to girls in the class?

   (a) 3:1   (b) 2:1   (c) 1:3   (d) 1:2

16. The percentage score of a student in 30 tests in different semesters is 54.66. His 'range' in scores (highest score - lowest score) is 80%. If these two tests are excluded, the average score is 55%. What is his highest score?

   (a) 77%   (b) 80%   (c) 85%   (d) 90%

17. A car covers 90kms in 3 hours, and the remaining 150 kms in 4 hours. What is the average speed?

   (a) 32.5 kmph   (c) 34.286 kmph
   (b) 33.33 kmph   (d) 35.123 kmph

Solutions

1. $11 + 22 + 33 + 44 + 55 + 66 + 77 + 88 + 99$ are the 2 digit nos sum $= 495$

   $\therefore$ 55 ans

2. $\dfrac{n(1+n)}{2} = 3n + 3$; (given)

   $\therefore n^2 + n = 6n + 6$

   $\Rightarrow n^2 - 5n - 6 = 0$

   $\Rightarrow (n - 6)(n + 1) = 0$

   $\therefore n = 6$

3. Let $a, a+1, a+2 \cdots\cdots a+8$ be the no. Their Av. is,

   $\therefore \dfrac{9a + 36}{9} = 24$ (given)

   $\therefore 9a + 36 = 216$

   $9a = 180$

$a = 180/9 = 20$

The no. are 20, 21, 22, 23, 24, 25, 26, 27, 28.

4th no. is 23

4. The odd no. s are 27, 29, 31, 33, 35 (Their average = 31) (given)

   Even number between 27 and 35 are 28, 30, 32, 34.

   Their average is $\dfrac{30 + 32}{2} = 31$

   (same as their $\dfrac{sum}{4} = \dfrac{124}{4} = 31$)

5. $\dfrac{2 \times 69 + 2 \times 34 + 2 \times 4}{2 + 2 + 2} = \dfrac{138 + 68 + 8}{6} = \dfrac{214}{6} = 35\dfrac{2}{3} y$

6. There are 5 Sundays 1st, 8th, 15th, 22nd, 29th; and 25 other days

   $\therefore$ Required average is $\dfrac{5 \times 48 + 25 \times 24}{30} = \dfrac{240 + 600}{30} = \dfrac{840}{30} = 28$

7. $\dfrac{60 \times 49 + 60 \times 45.5 + 62 \times 42}{60 + 60 + 62} = \dfrac{2940 + 2730 + 2604}{182}$

   $= \dfrac{8274}{182} = 45.46\%$

8.
	m		items
Total for 3m	12	×	3750 = 45000:
	3	×	3000 = 9000
For 9m			36000

   Av. 4000 per month.

9. 8 @ 3.5 Runs → 28: Total for 8 overs

   $\therefore$ 32 overs @ x run → 160 (since total for 40 overs = 188)

   $\therefore$ x = 5

Averages 85

10. $(a + b + c + d + e) = 5 \times 24 = 120$;

$a = \dfrac{(b + c + d + e)}{5}$ ∴ $a + b + c + d + e = ba = 120$

$a = 20$

11. $10a - 62 + x = 10(a + 1.5)$

∴ $x = 62 + 15 = 77$ kg

12. Let 'a' be the average in the 5 innings

$5a + 102 = 6(a + 4)$

∴ $a = 102 - 24 = 78$

New average = 82 runs

13. Let w be the wickets till then

$\dfrac{10.8w + 22}{w + 3} = 10.8 - 0.8 = 10$

$\Rightarrow 10.8w + 22 = 10w + 30$

∴ $0.8w = 8$

∴ $w = \dfrac{8}{8} \times 10 = 10$

14. Let small letter denote the total weight of $(p + q + r) = 3 \times 81$
    = 243 kgs;

    $p + q + r = s = 4 \times 75 = 300$ kg

    Then s = 57 kg; t = 69 kg;

    Given = $q + r + s + t = 4 \times 74 = 296$

    $\Rightarrow q + r + (57 + 69) = 296 \Rightarrow (q + r) = 170$ kg

    ∴ $p = 243 - 170 = 73$ kg

15. Let b : g be the ratio

    $(b + g) 70 = 49b + 77g$

    $\Rightarrow (b + g) 10 = 7b + 11g$

⇒ Dividing by g throughout,

$$10\frac{b}{g} + 10 = \frac{7b}{g} + 11 \Rightarrow \frac{3b}{g} = 1$$

$$\therefore \frac{b}{g} = \frac{1}{3}.$$

They are in the ratio of 1:3

Check   L.H.S $(1 + 3) 10 = 40$;

R.H.S $7 \times 1 + 11 \times 3 = 40$ QED

16. Let x be the highest score

    Lowest = x - 80

    Averages given 54.66 and 55 for 30 and 28 tests

    $$\frac{30 \times 54.66 - x - x + 80}{28 \times 55} = \frac{1639.8 - 2x + 80}{1540} = 1$$

    $\therefore$ 1639.8 - 1540 = 2x - 80

    $$\Rightarrow \begin{array}{r} 1639.8 \\ -1540 \\ \hline \simeq 100 \end{array} = 2x - 80$$

    $\therefore$ x $\simeq$ 90 (highest score); lowest score) 10

17. $\frac{90 + 150}{3 + 4}$ $\frac{240}{7}$ = 34.286 kmph

| Key |

1)	c	2)	b	3)	d	4)	c	5)	d	6)	b
7)	c	8)	c	9)	c	10)	d	11)	c	12)	d
13)	b	14)	c	15)	c	16)	d	17)	c		

12

Ages

Exercises

1. Suresh is younger than Ramesh by 2 years. The ratio of their ages is 9:10. What is the age of Ramesh?

   (a) 13   (b) 15   (c) 20   (d) 30

2. Anand's and Bhushan's ages are in the ratio 5:7. B is 8 years older than A. What will be the ratio of their ages 2 years later?

   (a) 2:3   (b) 11:15   (c) 3:4   (d) 5:7

3. A person's age after 10 years will be twice his age twenty years earlier. What is his age?

   (a) 27   (b) 38   (c) 45   (d) 50

4. The sum of the ages of A and B is 80 years. Their difference is 4 years. What are the ages?

   (a) 17, 51   (b) 19, 36   (c) 38, 12   (d) 42, 38

5. The sum of the ages of A and B is 80; Their product is 1596. What are their ages?

   (a) 17, 51   (b) 19, 36   (c) 38, 12   (d) 42, 38

6. The square of B'S age is 1444 years^2 (units). The LCM of the ages of A and B is 798 and their HCF = 2. What is the age of A?

   (a) 17years   (b) 19years   (c) 38years   (d) 42years

7. A is 4 years elder to B. The ratio of their ages is 21/19. What are their ages?

   (a) 17, 51   (c) 38, 12   (e) 84, 76
   (b) 19, 36   (d) 42, 38

8. Ages of A and B are in the ratio of 7:5. Three years later the ratio will be 4/3. What is the sum of the ages?

   (a) 27years   (b) 35years   (c) 36years   (d) 40years

9. Ages of A and B bear the ratio 7:5. After 10years A will be 31years. What is B's age now?

   (a) 14years   (b) 15years   (c) 25years   (d) 32 years

10. A'S and B'S ages are in the ratio 7:5. If the difference of B' present age and A'S age after 5 years is 11years. What is the product of the present ages?

    (a) $75y^2$ (b) $195y^2$ (c) $245y^2$ (d) $315y^2$

11. Gopal's age is 6/13 of his mother. After 5years the ratio will be 1/2. What is Gopal's age now?

    (a) 6years   (b) 13years   (c) 30years   (d) 65years

12. The ratio of the ages of a father and son is 4:1 in 1977. After 20 years the ratio is 2:1. When is the son born?.

    (a) 1957   (b) 1967   (c) 1977   (d) 1987

13. A person is 30 years elder to his son. In 5years more their ratio will be 3:1. What is the son's age now?

    (a) 5years　　(b) 10years　　(c) 15years　　(d) 20years

14. Teachers age: Students age is 2:1. After 20years the ratio is 3:2. What is the difference between their ages now?

    (a) 15　　(b) 20　　(c) 25　　(d) 30

15. The sum of the ages of 6kids formed at 4 yrs intervals is 66. What is their average?

    (a) 3　　(b) 7　　(c) 11　　(d) 15

16. Aruna's father was 38 years of age when she was born. Her mother was 36 years when her brother Gopi was born. Gopi is 4years younger than Aruna. What is the difference of the ages of the parents?

    (a) 2years　　(b) 4years　　(c) 6 years　　(d) 8years

Solution

1. $S + 2 = R;\ \dfrac{S}{R} = \dfrac{9}{10}\ \therefore 9R = 10S \Rightarrow 9(s+2) = 10s$

   $\therefore S = 18 \qquad R = 20$

2. $\dfrac{a}{b} = \dfrac{5}{7} \therefore 7a = 5b \therefore 7a = 5(a+8)$

   $\Rightarrow 2a = 40 \qquad a = 20;\qquad b = 28$

   Required $= \dfrac{20+2}{28+2} = \dfrac{22}{30} = \dfrac{11}{15}$

3. $(x + 10) = 2(x - 20) \Rightarrow 2x - x = 10 + 40$

   $\therefore x = 50$

4. $\underline{\begin{array}{l} a + b = 80 \\ a - b = 4 \end{array}}$

   Add $2a = 84$

$a = 42 \quad \therefore b = a - 4$
$\qquad\qquad = 42 - 4$
$\qquad\qquad = 38$

5. $a + b = 80; \quad ab = 1596$
$\therefore a(80 - a) = 1596$
$\therefore 80a - a^2 = 1596$
$\therefore a^2 - 80a + 1596 = 0$
$\Rightarrow a^2 - 42a - 38a + 1596 = 0$
$\Rightarrow a(a - 42) - 38(a - 42) = 0$
$\Rightarrow (a - 42)(a - 38) = 0$

$\left. \begin{array}{l} a = 42; \\ a = 38 \end{array} \right\} \left. \begin{array}{l} b = 38; \\ b = 42 \end{array} \right\}$

```
     1596
     /\
    2  798
       /\
      2  399
         /\
        2  133
           /\
          19  7
```

$pr = 1596$

1596	×	1
798	×	2
349	×	4
133	×	12
42	×	38

Divisibility by 7: In 1596 subtract 2×6 from 159
We get 147
Which divisible by 7

6. $b^2 = 1444 \qquad \therefore b = \sqrt{1444} = 38$
b = 38 years HCF of a, b is 2;
LCM is 798
Since a b = Their HCF × LCM
$\therefore a \times 38 = 2 \times 798$
$\therefore a = 42$ years

```
       14 44|38
          9
         ___
         544
      68 
         544
         ___
          x
```

7. $a = b + 4 \rightarrow$ (I)
$\dfrac{a}{b} = \dfrac{21}{19} \rightarrow (II)$
$\therefore \dfrac{b + 4}{b} = \dfrac{21}{19} \Rightarrow 19b + 76 = 21b$
$2b = 76$
$\therefore b = 38$ years;
$\therefore a = b + 4 = 42$ years

8. $\dfrac{a}{b} = \dfrac{7}{5}$; $\dfrac{a+3}{b+3} = \dfrac{4}{3}$ $\therefore 3a + 9 = 4b + 12$
or $3a - 4b = 3$ (ii)
$5a - 7b = 0 \Rightarrow 15a - 21b = 0$ (i)
(ii) $\Rightarrow 15a - 20b = 15$
Subtract $-b = -15$
$\therefore b = 15$ years $a = \dfrac{7}{5} \times 15 = 21$ year
$\therefore a + b = 36$ years

9. $\dfrac{a}{b} = \dfrac{7}{5}$; $a + 10 = 31$ years, $a = 21$ years
$B = \dfrac{5}{7} \times a = \dfrac{5}{7} \times 21 = 15$ years

10. $\dfrac{a}{b} = \dfrac{7}{5}$ $\therefore 5a - 7b = 0$
(i) $[(a+5) \sim b] = 11$ years
Since a's age is larger $\left(\dfrac{a}{b} \text{ being } \dfrac{7}{5}\right)$
$(a+5) - b = 11$
$\therefore a - b = 6$ (ii) $\Rightarrow 5a - 5b = 30$
(i) $\quad 5a - 7b = 0$
Subtract $2b = 30$
$\therefore b = 15, a = 21$
$A = \dfrac{7}{5} \times 15 = 21$ years
Required product $= 21 \times 15 = 315$ years

11. Let Gopal's age $= g$, Mother's age $= m$. say
$\dfrac{g}{m} = \dfrac{6}{13}$ (i); $\dfrac{g+5}{m+5} = \dfrac{1}{2}$ (ii)
$13g - 6m = 0$ (i)
$2g + 10 = m + 5$ (ii)
From (i) $13g - 6m = 0$
$6m - 12g = 30 \quad \Rightarrow [\because m - 2g = 5$ (ii)]

92 Quantitative Aptitude

add g = 30 ans $m = \dfrac{13g}{6} = 13 \times \dfrac{30}{6} = 65$ years

Check $\dfrac{30+5}{65+5} = \dfrac{35}{70} = \dfrac{1}{2}$ QED

12. Let f_1, s_1, be the ages in 1977;

f_2, s_2 in 1997

$\dfrac{f_1}{s_1} = 4$ (i); $\dfrac{f_2}{s_2} = 2 \Rightarrow \dfrac{f_1+20}{s_1+20} = 2$ or

$f_1 - 2s_1 = 20;$

f_1 from = 40 years the son is born 10 years before 1977 i.e., 1967 (since $= s_1 = 10y$ in 1977) ans

Details:

$f_1 - 2s_1 = 20$ (ii)

$\underline{f_1 - 4s_1 = 0}$ (i)

Subtract $2s_1 = 20$

$s_1 = 10$ years

13. $f = s + 30$ (i) ; $\dfrac{(f+5)}{(s+5)} = 3 \Rightarrow f + 5 = 3s + 15$ (ii)
    $f = 3s + 10$ (ii)

Subtract $2s - 20 = 0$ $f = 40$

$s = 10$ years ans

Check $\dfrac{45}{15} = 3$ QED

$\therefore s = 10$ years

14. $\dfrac{T}{s} = \dfrac{2}{1}$ (i); $\dfrac{T+20}{s+20} = \dfrac{3}{2}$ (ii)

$\Rightarrow 2T + 40 = 3s + 60 \Rightarrow 3s - 2T = -20$ (ii)

$2T - 4s = 0$ (i) $\quad\quad\quad 2T - 4s = 0 (i)$

$\quad\quad\quad\quad\quad\quad\quad\quad\quad$ Add : - s = - 20

Required $T - s = 40 - 20$ | $s = 20$
$\quad\quad\quad\quad = 20$ ans | $T = 40$

15. $(x - 8) + (x - 4) + (x) + (x + 4) + (x + 8) + (x + 12) = 66$

$6x + 12 = 66;$

$6x = 54;$

$x = 9$

Their ages are 1, 5, 9, 13, 17, 21

Their average is $\dfrac{66}{6} = 11$ years ans

16.

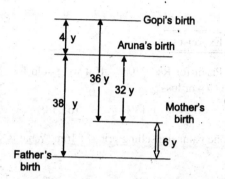

From the figure; the required difference of ages of parents (shown in double line) is 6 years

Ans. (c)

Key

1)	c	2)	b	3)	d	4)	d	5)	d	6)	d
7)	d	8)	c	9)	b	10)	d	11)	c	12)	b
13)	b	14)	b	15)	c	16)	c				

13

Profit and Loss

Exercises

1. A radio was sold at 15% Profit for Rs. 2300/-. If it were sold for Rs. 2260/- What would be % p or loss?

    (a) 9%   (b) 11%   (c) 13%   (d) 14%

2. By selling 48 m of cloth, one gains the selling price of 16m. What is the % gain?

    (a) 16%   (b) 32%   (c) 48%   (d) 50%

3. A retailer marks his items at 40% more than cost price and then gives a discount of 20% on the M.R.P. What is the actual profit on these items?

    (a) 35%   (b) 30%   (c) 20%   (d) 12%

4. A producer allows a commission of 25% on the retail price and earns a profit of 7.5%. What would be the profit percentage if the commission is reduced by 10%?

    (a) 15.45%   (b) 16.6%   (c) 18.7%   (d) 21.83%

5. Gopal purchased a second hand two wheeler for Rs. 15,000/-; and spends Rs. 3000/- for repairs? He sells it for Rs. 20,000/- what is the percent profit?

    (a) 10%   (b) $11\frac{1}{9}\%$ (c) $12\frac{1}{2}\%$ (d) 15%

6. A middle person purchases onions at the government subsidized price of Rs. 7/- and sells at Rs. 12/- per kg in the scarce days. What is the profit percent per 20 kg?

    (a) 58.95%   (b) 60.67%   (c) 65.33%   (d) 71.43%

7. A person purchased toys @ Rs. 360/- per dozen and sells each one at Rs. 32/-. What is the percent profit?

   (a) 3.33%   (b) 6.67%   (c) 10%   (d) 12.5%

8. Coconuts are purchased @ Rs. 500/- per hundred and sold at Rs. 84/- per dozen. What is the percent profit?

   (a) 25%   (b) 32%   (c) 40%   (d) 45%

9. Suresh purchased a cycle for Rs. 1800/- and sold at a loss of 10%. What is the SP?

   (a) Rs. 1475/-   (b) Rs. 1525/-   (c) Rs. 1580/-   (d) Rs. 1620/-

10. A person gains 25% by selling an article for a certain price. If he sells it at double the price; the percentage profit will be

    (a) 50   (b) 75   (c) 100   (d) 125   (e) 150

11. In a store, profit is 300% of cost. If the cost is increased by 25% but selling price remains constant, What percentage of SP is the profit?

    (a) 65.25%   (b) 68.75%   (c) 70%   (d) 71.5%   (e) 72.5%

12. The profit earned by selling an article for Rs. 732/- equals the loss incurred when it is sold at Rs. 348/- At what price it should be sold to make 20% profit?

    (a) Rs. 618/-   (c) Rs. 638/-   (e) Rs. 658/-
    (b) Rs. 628/-   (d) Rs. 648/-

13. If the CP of 150 pens equals the SP of 100 pens. What is the gain percentage?

    (a) 20%   (b) 25%   (c) 40%   (d) 50%   (e) 60%

14. Applied to a bill of Rs. 10,000/- the difference between a discount of 30% and two successive discounts of 25% and 5% is

    (a) Rs. 100/-   (c) Rs. 140/-   (e) Rs. 160/-
    (b) Rs. 125/-   (d) Rs. 150/-

15. Marked price is 25% higher. Two successive discounts of 10% are given on the marked price. What is the net profit percentage?

    (a) 0.25%   (b) 1.25%   (c) 2.25%   (d) 3.5%   (e) 6.5%

16. A trader marks an item 15% above CP. He gives a discount and suffers a loss of 1%. What is the discount allowed?

    (a) 9.875%   (c) 11.675%   (e) 13.913%
    (b) 10.935% (d) 12.825%

Solutions

1. | CP | P | SP | CP |
    |-----|-----|-----|-----|
    | 100 | 15% | 115 | → 100 |

    $2300 \to \dfrac{100}{115} \times 2300 = $ Rs. 2000/-

CP	P	SP	P
2000	?	2260	Rs. 260/-

    $\% P = \dfrac{260}{2000} \times 100 = 13\%$

2. Let CP of the cloth be Rs. 100/- per metre. Then 48m of cloth costs Rs. 4800/- Gain = SP of 16m. If x is the SP per m, CP of 48 m being 4800;

    SP will be $48 m \times \dfrac{xy}{m} = 48x$

    Gain = 1 SP of 16m (if x is the SP per metre)

    Gain = 48 x - 4800

    This is SP. of 16m ie. 16 x

    ∴ 48 x - 16x = 4800

    ∴ $x = \dfrac{4800}{32} = $ Rs. 150% per metre

    Gain = 48 × 150 - 48 × 100 = 48 × 50

    % gain = $\dfrac{48 \times 50}{4800} \times 100 = 50\%$

3. CP = Rs. 100/- MRP Rs. 140/-. Discount 20% on MRP or $\frac{20}{100}$ of 140

$= \frac{140}{5} =$ Rs. 28/-

$\therefore$ SP = $\left.\begin{array}{r} 140 \\ = \frac{28}{112} \end{array}\right\}$; net profit = 112 - 110 $\doteq$ Rs. 12/- or 12%

4. Let the cost price be Rs. x/- or $\frac{100x}{100}$

Retail price is different. Let the R.P be $\frac{rx}{100}$

Commission on RP is $\left(\frac{25}{100}\right)\left(\frac{rx}{100}\right)$

Sales price $= \frac{rx}{100}\left[1 - \frac{25}{100}\right]$;

Profit = SP - CP $= \frac{rx}{100}\left[1 - \frac{25}{100}\right] - \frac{100x}{100}$

This is given as $\frac{7.5}{100} x$

Where $[CP(x)] \Rightarrow \frac{x}{100}\left[r\left(\frac{75}{100}\right) - 100\right] = 7.5 \frac{x}{100}$

$\therefore$ 75 r - 100^2 = 750 $\Rightarrow$ 75 r = 10750

$\therefore$ r = $\frac{10750}{75} = \frac{2150}{15} = \frac{430}{3} = 143.3$

Thus CP = $x \equiv \frac{100x}{100}$ and RP = 143.3 $\frac{x}{100}$

— II transaction commission reduced by 10% i.e comm.

=15% $SP_2 = \frac{rx}{100} - \frac{15x}{100^2} = \frac{rx}{100}\left[1 - \frac{15}{100}\right] = \frac{85}{100^2} rx$

$$\text{Profit} = SP_2 - CP = \frac{85rx}{100^2} - \frac{100x}{100}$$

$$\Rightarrow \frac{x}{100}\left[\frac{85r}{100} - 100\right] - \frac{x}{100}\left[\frac{85 \times 143.3}{100} - 100\right]$$

$$\Rightarrow \frac{x}{100}\left[\frac{85 \times 143.3 - 100^2}{100}\right]$$

$$\text{Profit \%} = \frac{x}{100}\left[\frac{85 \times 143.3 - 100^2}{100}\right]\frac{100x}{100}$$

$$\frac{12183.3 - 10000}{100} = \frac{21,83.3}{100}x = 21.83\%$$

5. Total CP = 15,000   SP = 20,000 profit = 2000
   $$\begin{array}{r}+ 3,000 \\ \hline 18,000\end{array}$$

   Profit percent = $\dfrac{2000}{18000} \times 100 = \dfrac{100}{9} = 11\dfrac{1}{9}\%$

6. CP for 20 kg of onions = 20 × 7 = Rs. 140/-

   SP 20 × 12 = 240, profit = Rs. 100/-

   Profit percentage = $\dfrac{100}{140} \times 100 = \dfrac{500}{7} = 71.43\%$

7. CP per each = $\dfrac{360}{12}$ = Rs. 30/- SP Rs. 32/- per each

   Profit percent = $\dfrac{2}{30} \times 100 = \dfrac{20}{3} = 6.67\%$

8. CP Rs. 5/- each; SP = $\dfrac{84}{12}$ = Rs. 7/- per each

   Percent profit = $\dfrac{2}{5} \times 100 = 40\%$

9. CP = Rs. 1800/- loss 10% or Rs. 180/-

   $$SP = \begin{array}{r} 1800 \\ -\ 180 \\ \hline \text{Rs. } 1620 \end{array}$$

10. CP = x, profit = 25%

    $SP = x + \dfrac{x}{4} = \dfrac{5x}{4}$

    If the SP were to be $\dfrac{10x}{4}$; profit = $\dfrac{10x}{4} - x = \dfrac{6x}{4}$

    Profit % = $\dfrac{6x}{4x} \times 100 = \dfrac{600}{4} = 150\%$

11. SP = CP + profit = $\dfrac{100x}{300} + \dfrac{300x}{100} = \dfrac{400x}{100}$

    If CP is $\dfrac{125x}{100}$, P $\dfrac{400x}{100} - \dfrac{125x}{100} = \dfrac{275x}{100}$

    This will be in terms of SP

    $\left\{ \because SP \to \dfrac{400x}{100} \right\}$ $P \Rightarrow \dfrac{\dfrac{275x}{100}}{\dfrac{400x}{100}} \times 100$

    $= \dfrac{275}{400} \times 100 = \dfrac{275}{4}$

    $= 68.75\%$

12. CP = x, $SP_1 = 732$

    Profit = 732 - x; $SP_2 = 348$

    Loss = x - 348

    Given = 732 - x = x - 348

    ∴ 2x = 732

    +348

    1080

    ∴ x = Rs. 540/-

    100 CP → 120 SP

    540 CP → $\dfrac{120}{100} \times 540 = \dfrac{6}{5} \times 540$ = Rs. 648/-

13. Let c, s be CP and SP of each pen; $150c = 100s$

   $\therefore s = \dfrac{3}{2}c$; gain $= \dfrac{3}{2}c - c = \dfrac{c}{2}$

   Gain% $= \dfrac{c}{2} \cdot \dfrac{1}{c} \times 100 = 50\%$

14. Case (a) discount of 30%,

   Discounted amount $= \dfrac{70}{100} \times 10,000 = 7000$

   Case (b)

   (i) discount of 25%, discounted amount $= \dfrac{75}{100} \times 10,000 = 7500$;

   (ii) discount of 5%, discounted amount $= \dfrac{95}{100} \times 7500 = 7125$

   Difference $\therefore (7000 \sim 7125) =$ Rs. 125/-

15. CP 100 → MRP 125

   (i) 100 → 90   (1st discount)

   $125 \to \dfrac{18}{20} \times 125 = \dfrac{9}{2} \times 25$

   (ii) 100 → 90   (2nd discount)

   $\dfrac{9 \times 25}{2} \to \dfrac{18}{20} \times \dfrac{9 \times 25}{2} = \dfrac{405}{4} = 101.25$

   Net profit is **1.25 percentage**

16. CP = 100; MRP = 115; let discount be x% on MRP.

   If MRP is 100 → DP = 100 - x

   For MRP 115 → DP is → $\dfrac{100 - x}{100} \times 115 = \left(1 - \dfrac{x}{100}\right) 115$

   Loss% $= \dfrac{100 - \left(1 - \dfrac{x}{100}\right) 115}{100} \, 100\%$

   $\Rightarrow \left(100 - \left[\dfrac{x}{100}\right]\right) 115 = 1$ (given)

$\Rightarrow$ $100 - 115 + \dfrac{23}{20}x = 1$ $\therefore \dfrac{23}{20}x = 16$

$\therefore x = \dfrac{320}{23} = 13.9130$ percent

Check MRP = 115   Discount on MRP 13.9130

$DP = \dfrac{100 - 13.9130}{100} \times 115 = \dfrac{86.0870}{100} \times 115 = 99.0001$

Loss = 100 - 99.00001 $\simeq$ 1% QED

1)	c	2)	d	3)	d	4)	d	5)	b	6)	d
7)	b	8)	c	9)	d	10)	c	11)	b	12)	d
13)	d	14)	b	15)	b	16)	e				

14

Business Mathematics - Partnership (Joint Ventures)

Exercises

1. Gopal and Hemanth started a joint business with 80k, and 120k in January 2000 and after 4 months respectively. At the end of 2001, what is the difference in the shares of Gopal & Hemanth in the profit of 81 k?

   (a) 3 k   (b) 6 k   (c) 9 k   (d) 12 k   (e) 15 k

2. Narendhra and Obul Reddy started with 45k & 35k a joint venture in 2003. At the end of 3 years, what is the ratio of their profits?

   (a) 3 : 2   (b) 9 : 7   (c) 9 : 8   (d) 9 : 11   (e) 3 : 4

3. Rao, Sastry and Tarun started with 80k, 60k, 90k a joint venture. After 8 months Tarun left. If at the end of 10 months, the profit was Rs. 84,800/-. What is Tarun's share?

   (a) 26,900   (b) 27,800   (c) 28,000   (d) 28,400   (e) 28,800

4. Omkar and Paul invested in a business in the ratio 5: 7. Paul's and Roy's investments bear a ratio of 11: 13. If the profit at the end of the year is Rs.44,600/-. What is the share of Paul?

   (a) 11,000   (c) 18,200   (e) 20,000
   (b) 15,400   (d) 19,800

5. Saxena and Tarun started a venture with investments in the ratio 5: 7. After 1 year Uttam joined. Tarun and Uttam's investments are in the ratio 2 : 3. At the end of 2 years from the beginning in what ratio they should distribute their profits?

(a) 18: 25: 21   (c) 20: 28: 21   (e) 24: 31: 23
(b) 19: 26: 22   (d) 22: 30: 23

6. Anil, Biswas and Churchill invest in a venture in the ratio is $\frac{5}{2} : \frac{3}{2} : \frac{7}{3}$. After 3 months A and C withdraw 1/5th & 1/7th of their investments. At the end of an year from the beginning, profit is Rs. 82,200/-. What are their shares?

(a) 29,000, 29,000, 24,200
(b) 29,000, 30,000, 23,200
(c) 30,000, 30,000, 22,200
(d) 30,600, 21,600, 30,000
(e) 31,600, 20,600, 30,000

7. Arun and Prasad started a venture with investments in the ratio 3 : 4. After 4 months Arun withdrew 1/3 of his capital and Prasad withdrew 1/4th of the capital. The profit at the end of the year is Rs.5,100/-. What is the difference of the profits?

(a) Rs.850/-   (c) Rs.920/-   (e) Rs.1090/-
(b) Rs.900/-   (d) Rs.1,030/-

8. Murthy, Naveen and Paul partnered for 12m, 10m & 8 m respectively. At the end of the year they shared the profits in the ratio of 2: 3: 4. What was the ratio of their investments?

(a) 5: 8 : 11   (c) 6 : 9 : 15   (e) 7:12:14
(b) 5 : 9 : 15   (d) 6:10:11

9. Anusha and Esha invest in the ratio 4 : 3 in a business. At the end of the year they give 2% of the profits for charity and Esha gets Rs. 8,400/-. What is the total profit?

(a) 14,000   (b) 18,000   (c) 19,000   (d) 20,000   (e) 21,000

10. Pradip, Qureshi and Roy start a joint venture. Three times Pradip's investment equals four times Qureshi's and Qureshi's investment is thrice that of Roy. If the total profit at the end of the year is 32k. What is the share of Roy?

   (a) 4k   (b) 12k   (c) 16k   (d) 20k   (e) 32k

11. If Syam and Tarak shared their annual profit in the ratio 5 : 6 and if Syam's investment is 2.5 k. What was the total investment?

   (a) 5,300   (b) 5,500   (c) 5,700   (d) 5,900   (e) 5,950

12. A summer cottage was booked by three families for 8, 10, 12 days in June. The monthly rent was Rs.30,000/-. What is the amount paid by the 1st one?

   (a) 8,000   (b) 10,000   (c) 12,000   (d) 14,000   (e) 16,000

13. A invests 1/4th of the capital for 1/4th time. B invests 1/3 of the capital for 1/3 time and C invests the remaining amount for the whole time. What is the share of A if the total profit at the end was 74k?

   (a) 17,200   (b) 17,500   (c) 18,000   (d) 18,600   (e) 19,300

14. Investment of P is 4 times that of Q. The time for which P kept the amount is 3 times that of Q. If the total profit is 39k. What is that of P?

   (a) 30k   (b) 31k   (c) 35k   (d) 36k   (e) 38k

Solutions

1.

	G	H
2000 Jan	80 k	—
After 4 m		120 k
End of the 2001 (2y)	↓2y	↓5/3y
Profit 36 k	160	200

Ratio = 160 : 200

G's share = $\frac{4}{9} \times 81k = 36k$

H's share = 45 k

Difference = 9 k

Equivalent investments

2.

Time	K	L
2003	45 k	35 k
2004	↓	↓
2005		
Profit	9 :	7

Profit ratio = 45: 35 = 9: 7

Answer is (b)

3. Ratio of equivalent investments

   R: S : T = 80 k × 10 m;   60 k × 10 m ;   90 k × 8m

       = 800   ;   600   ;   720

       = 8   ;   6   ;   7.2

   Total = 21.2

   Tarun's share $= \dfrac{7.2}{21.2} \times 84800 = 28,800$ **Answer is: (e)**

Time	R	S	T
0	80 k	60 k	90 k
End of 8 months	10m	10m	8m
End of 10 months	↓	↓	
Profit 84.8 k	8	6	7.2

4.   Omkar: Paul = 5: 7;

                        Paul : Roy   = 11 : 13   LCM of 7 & 11 = 77

   Share of profits : Omkar : Paul : Roy = 55 : 77 : 91

                            (Total 223)

   Share of Paul $= \dfrac{77}{223} \times 44600 = 15,400$

                      (L.C.M of 7, 11 being 77)

   Shares: O:11,000 ; R : 18,200; P : 15,400 = **44,600** (Total)   **(b)**

5.

Time ↓	S	T	U

0
12 m
24 m
	2y ⎰ 5x
	     ⎱ or
	       ↓10x
		7x ⎱ 2y
		14x ↓
			21x
			↓ 1y
			↓

Ratio of equivalent Investments = Ratio of Profit's	20x 20	28x 28	21x 21

Let 5x, 7x be the investments of Saxena and Tarun; Tarun's and Uttam's $2 : 3 \Rightarrow 14 : 21$; $S : T = 10x : 14x$. Total = 69 Required ratio: 20 : 28 : 21   (c)

6. Investments of A : B : C $= \dfrac{5}{2} : \dfrac{3}{2} : \dfrac{7}{3} = \dfrac{15 : 9 : 14}{6}$; say $15x : 9x : 14x$

After 3 m investments are $(15x - 3x); 9x;$ and $(14x - 2x)$ i.e

$$12x : 9x : 12x$$

Equivalent investments

$= (15x \times 3m + 12x \times 9m) : [(9x \times 12m)] : (14x \times 3m + 12x \times 9m)$

$= (45x + 108x) : (108x) : (42x + 108x)$

Time	A	B	C
0	15x ↓ 3m	9x ↓ 3m	14x ↓ 3m
3m	12x ↓ 9m	9x ↓ 9x ↓ 9m	12x ↓ 9m
12m	↓	↓	↓
Total profit Rs. 82,000/-	153	108	150 Total 411
Equivalent Shares	30,600	21,600	30,000

$= 153x : 108x : 150x$
Total $= 411$; T.P : 82,200
Shares: $\dfrac{153}{411} \times 82{,}200 = 30{,}60$
$\dfrac{108}{411} \times 82200 = 21{,}600$
and $\dfrac{150}{411} \times 82{,}200 = 30{,}000$
Total : 82,200

7.

Time	A	P
0	3x	4x
4m	↓ 4m	↓ 4m
	2x	3x
12m	↓ 8m	↓ 8m
Total profit Rs. 5,100/-	7 :	10 :
Shares of Equivalent inv. Shares of Profit	7	

Equivalent investment

$A : P = (3x \times 4m + 2x \times 8m)$
$= (4x \times 4m + 3x \times 8m)$
$= (12x + 16x) : (16x + 24m)$
$= 28x : 40x = 7 : 10$

17

$\dfrac{7}{17} \times 5100 = 2100$ and 3000

$= (3000 - 2100) = Rs.900/-$ (b)

8.

m, n, p be the investments

Ratio of Equivalent investments
$= 12m : 10n : 8p$
$=$ Ratio of Profits $= 2 : 3 : 4$

$\therefore \dfrac{12m}{10n} = \dfrac{2}{3}$ and $\dfrac{10n}{8p} = \dfrac{3}{4}$

$\therefore 36m = 20n;\ 40n = 24p$

$\therefore m : n = 5 : 9$

Eq = 12m : 10n : 8P
Profit = 12m : 10n : 8p
$= 2 : 3 : 4 = 1 : 1.5 : 1.5$

$\therefore \dfrac{m}{n} = \dfrac{20}{36} = \dfrac{5}{9}$; and $\dfrac{n}{p} = \dfrac{24}{40} = \dfrac{3}{5}$ $\therefore n : p = 3 : 5$ or $9 : 15$

$\therefore m : n : p = 5 : 9 : 15$ (b)

9.

Time ↓	A	E
0 m	4x	3x
Ratio of Profits	4	3

Let $4x, 3x$ be the Iv. If 100 units is profit 2% is charity and $\dfrac{98}{100}$ is shared in the ratio of $4x : 3x$ or $4 : 3$.

$\therefore \dfrac{3}{7} \times \dfrac{98}{100} \times 100U = Rs.\ 8400$ (given)

$\therefore 100U = 8400 \times \dfrac{7}{3} \times \dfrac{100}{98} = 20,000$

10.

Time ↓	P	Q	R
0 m	p	q	r
	p	3/4p	p/4

Let p, q, r be the investments:
Given $3p = 4q$ and $q = 3r \Rightarrow q = (3/4)$ p and $r = q/3 = p/4$.

Ratio of the profits $= p : 3/4p : p/4 = \dfrac{4p : 3p : p}{4}$

i.e ratio is 4 : 3 : 1 & total = 8.

∴ R's share : $\dfrac{1}{8}$th of total $= \dfrac{1}{8} \times 32k = 4k$ (a)

11.

Time	S	Q
0 m	5U	6U
Profit	5	6

Let 5p, 6p be the profits. Total profit = 11p
investment = 5U : 6U. Total investment = 11U

Share of profit of Syam $= \dfrac{5}{11}$ = share of Inv of Syam = 2.5 k.

∴ Total Inv. $= 2.5k \times \dfrac{11}{5} = \dfrac{11}{2}k = 5.5k = 5500$ (with 0.5 above 2.5 and 5)

Inv. of T $= \dfrac{6}{5} \times 2.5k = 3k = 3000$ (b)

12. $\dfrac{8}{8+10+12} \times 30{,}000 = \dfrac{8}{30} \times 30{,}000 = 8000$ (a)

13. Let C be the total capital: A puts in C/4 rs for 1/4th time. (say 1/4 year)
B: C/3 for 1/3 of year and $C\left(1 - \dfrac{1}{4} - \dfrac{1}{3}\right) C = \dfrac{12-3-4}{12}C = \dfrac{5}{12}C$ for the whole year ∴ Equivalent investment

$= \left(\left(\dfrac{C}{4} \times \dfrac{1y}{4} : \dfrac{C}{3} \times \dfrac{1y}{3}\right) : \left(\dfrac{5}{12}C\right) : 1y\right) = \left(\dfrac{C}{16} : \dfrac{C}{9} : \dfrac{5C}{12}\right)$

$= C \dfrac{9 : 16 : 12}{144}$; 9 : 16 : 12

37. A's share $= \dfrac{9}{37} \times 74000 = 18000$ (c)

investments p = 4q; Equivalent investment
3 : q(1). Ratio of profits = 12q : q = 12 : 1

Profit of p $= \dfrac{12}{13} \times 39k = 36k$ (d)

Business Mathematics - Partnership (Joint Ventures) 109

Key

1)	c	2)	b	3)	e	4)	b	5)	c	6)	d
7)	b	8)	b	9)	d	10)	a	11)	b	12)	a
13)	c	14)	d								

15

Direct and Indirect Proportion

> **Exercises**

1. In a hostel, provisions were purchased at the beginning of the month (30 days) for 250 students. After 12 days, the inmates have increased to 300. For how many days will the provision lost?

   (a) 15   (b) 16   (c) 17   (d) 18   (e) 19

2. Four machines in one minute can reproduce 200 copies of a sheet. How many copies can be reproduced by ten machines in 20 minutes?

   (a) 7900   (b) 8500   (c) 9000   (d) 9050   (e) 10,000

3. Subhash can read 50 pages per hour. How many pages can be read in 40 minutes?

   (a) 29 1/2   (b) 33 1/3   (c) 36 2/3   (d) 37.8   (e) 40

4. If 1/4 kg potatoes cost Rs 5/2, what is the price of 5/2 kg?

   (a) Rs. 5/-   (b) Rs. 10/-   (c) Rs. 15/-   (d) Rs. 20/-   (e) Rs. 25/-

5. A college adopts a dress code. Cloth is purchased which will suffice for 60 teachers (or) 80 students. Material is used already for 45 teachers. For how many students will the remaining material be enough?

   (a) 5s   (b) 10s   (c) 15s   (d) 20s   (e) 25s

6. Cost of four 35ml packs is Rs 180/- what is the cost of 3packs of 70ml packs?

   (a) 90   (b) 135   (c) 180   (d) 240   (e) 270

7. A wheel with 8 cogs is meshed with a large wheel with 12 cogs. When the smaller wheel makes 30 revolutions, what is the number of revolutions made by the larger wheel?

   (a) 15   (b) 20   (c) 30   (d) 40   (e) 45

8. If a kg of capsicum i.e Rs.24/-, how many kg of capsicum can a person purchase for Rs100/-?

   (a) 3.655kg (b) 3.72kg (c) 3.95kg (d) 4kg (e) 4.166kg

9. In a land map 1 cm represents 25.4 inches, what is the actual distance between two points; whose length in the map is 70.2cm?

   (a) 3.82   (c) 4.362   (e) 49.53
   (b) 4.153   (d) 4.5750

10. A hospital needs 150 pills to treat 6 patients for a week. How many pills does it need to treat 10 patients for a weak?

    (a) 90   (b) 180   (c) 200   (d) 250   (e) 300

11. If 500 pounds of mush will feed 20 pigs for a week, for how many days will 200 pounds of mush feed 14 pigs?

    (a) 4 days   (b) 5 days   (c) 6 days   (d) 7 days   (e) 8 days

12. A road can be repaired in 20days by 16persons, working 6hr/day. In how many days can the same work be done. If 32 persons work 8 hours / day?

    (a) $6\frac{1}{2}$ (b) $6\frac{4}{5}$ (c) $7\frac{1}{2}$ (d) $8\frac{1}{4}$ (e) $8\frac{3}{4}$

Quantitative Aptitude

13. If 6men working 6 hr./day for 6days can complete 6 units of work in a 'Graemeena upadhi padhakam', how many units can 12 persons complete in 12 days working for 9 hours per day?

    (a) 24   (b) 28   (c) 30   (d) 32   (e) 36

14. 18 labours dig a ditch 36m long in 20days working 8hr/day. What is the additional no. of labours required to dig a trench 54m long, in 10days working 9 hr/day?

    (a) 25   (b) 30   (c) 35   (d) 40   (e) 45

15. Assembly of a certain number of cars can be completed by a group of technicians, in 90 days. If five of them fall sick, the work takes 5 days. What is the initial number of technicians?

    (a) 83   (b) 86   (c) 89   (d) 92   (e) 95

Solutions

1. Provisions necessary for (30 - 12) ie 18days, are available for 250 students. More students → less days (indirect proportion)

    $\therefore 300 : 250 = 18 : x$

    $\therefore 300x = 250 \times 18$

    $\therefore x = \dfrac{250 \times 18}{300} = \dfrac{5}{6} \times 18 = 15$ days

2. More machines more copies **DP**

    More time more copies **DP**

    $\left. \begin{array}{l} \text{Machines } 4:10 \\ \text{Time (min) } 1:20 \end{array} \right\} = 200 : x$

    $\therefore 4 \times 1 \times x = 200 \times 10 \times 20$

    $\therefore x = \dfrac{200 \times 10 \times 20}{4 \times 1} = 10,000$ copies

3. More time → more pages DP

   $\therefore$ 60min : 40min = 50P : xp

   $\therefore$ $60x = 40 \times 50$

   $\therefore$ $x = \dfrac{40 \times 50}{60} = 33\dfrac{1}{3}$ P

4. More quantity → more price DP

   $\therefore$ kg $= \dfrac{1}{4} : \dfrac{5}{2} = \dfrac{5}{2} : r$ Rs

   $\therefore$ $\dfrac{1}{4}r = \dfrac{25}{4}$

   $\therefore$ $r = 25/-$

   $r =$ Rs.25/-

5. More material → more people (teachers / students) D.P. 45 teacher's dress already prepared. Remaining material will be sufficient for 15 teachers more. If this material is to be used for students, cloth for 60 teachers = cloth for 80 students: cloth for 1 teacher = cloth for $\dfrac{90}{60}$ student; cloth for 45 teachers = $\dfrac{80}{60} \times 45$

   $= \dfrac{80 \times 3}{4} = 60$ cloth for 60 students

   $\therefore$ cloth for $15T = \dfrac{80}{60} \times 15$ cloths for 20 s.

   DP total 80 : n = total 60 : 15 (available for)

   Student std T : T

   $\therefore$ $60n = 80 \times 15$

   $\therefore$ $n = \dfrac{80 \times 15}{60} = \dfrac{4}{3} \times 15 = 20$ students

6. More volume (content) per pack more price DP

   More packs → more price DP

   $\therefore$ $\left.\begin{array}{l} \text{35ml; 70ml} \\ \text{4 pack; 3packs} \end{array}\right\} = 180 : x$ rs.

$$\therefore \frac{70 \times 3 \times 180}{35 \times 4} = 270 \text{ ans}$$

$(\because 35 \times 4 \times x = 70 \times 3 \times 180)$

7. The smaller wheel with less number of cogs: the larger the no. of rev (I. P)

   no. of cogs rev.

   $\therefore 12 : 8 = 30 : x$

   $\therefore 12x = 8 \times 30$

   $\therefore x = \dfrac{8 \times 30}{12} = \dfrac{240}{12} = 20 \text{ rev.}$

                       Qty.   price

8. More qty → more price DP : 1kg : x kg = 24 : 100

   $\therefore x = \dfrac{1 \times 100}{24} = 4.166 \text{kg}$

9. Longer distance in the scaled map longer the actual distance DP

   $\therefore 1 : 70.2 = 25.4 : l$

       cm   inches

   $\therefore l \times 1 = 70.2 \times 25.4 \text{ in}$

   $= \dfrac{70.2 \times 25.4}{12 \times 3} = 49.53 \text{ yards}$

10. More patients need more pills. DP

    $\therefore 6 : 10 = 150 : P$

    Patient   pills

    $\therefore 6p = 150 \times 10$

    $\therefore P = \dfrac{150 \times 10}{6} = 250 \text{ pills}$

11. More pigs consume in less days IP

    More mush lasts for more days DP

    $\therefore \left.\begin{array}{l} 14 : 20 \text{ pigs} \\ 500 : 200 \text{ lb mush} \end{array}\right\} 7 : d \text{ days}$

$\therefore 14 \times 500 \times d = 20 \times 200 \times 7$

$\therefore d = \dfrac{20 \times 200 \times 7}{14 \times 500} = 4$ days

12. More persons → less days IP

    More hours per day → less days IP

    $\therefore \left.\begin{array}{l} 32:16 \text{ persons} \\ 8:6 \text{ (hours/ day)} \end{array}\right\} = 20 : d \text{ days}$

    $\therefore 32 \times 8 \times d = 16 \times 6 \times 20$

    $\therefore d = \dfrac{16 \times 6 \times 20}{32 \times 8} = \dfrac{60}{8} = \dfrac{15}{2} = 7\dfrac{1}{2}$ days

13. $\left.\begin{array}{l}\text{More men → more units of work DP} \\ \text{More hr/day → more units DP} \\ \text{More days → more units of work DP} \\ 6 \times 6 \times 6 \times u = 12 \times 9 \times 12 \times 6\end{array}\right\} \left.\begin{array}{l}\text{men } 6:12 \\ \text{hr per day } 6:9 \\ \text{days } 6:12\end{array}\right\} 6:U$

    $\therefore U = \dfrac{12 \times 9 \times 12 \times 6}{6 \times 6 \times 6} = 36$ units of work

14. $\left.\begin{array}{l}\text{Longer the ditch → more labours DP} \\ \text{More no. of days → less labours IP} \\ \text{More hours per day → less laborers IP} \\ \therefore 36 \times 10 \times 9 \times l = 54 \times 20 \times 8 \times 18\end{array}\right\} \left.\begin{array}{l}\text{length } 36:54 \\ \text{days } 10:20 \\ \text{hr/day } 9:8\end{array}\right\} 6:U$

    $l = \dfrac{54 \times 20 \times 8 \times 18}{36 \times 10 \times 9} = 48$

    Additional no. required = 48 - 18 = 30 technicians

15. More technicians → less days IP $(t - 5: t) = 90:95$ days

    $\therefore (t - 5)95 = 90t$

    $\therefore 5t = 95 \times 5 \quad \therefore t = 95$

Key

1)	a	2)	e	3)	b	4)	e	5)	d	6)	e
7)	b	8)	e	9)	e	10)	d	11)	a	12)	c
13)	e	14)	b	15)	e						

16

Time and Work

Exercises

1. A man and woman can do a job in 5 days. The man alone can do in 8 days. How much time does the woman take to do the job alone?

   (a) $8\frac{1}{2}$ days (c) 10.2 days (e) $13\frac{1}{3}$ days
   (b) $9\frac{1}{4}$ days (d) 11.9 days

2. x, y, z take 28, 14 and 7 days separately to complete a work together. How much time do they take?

   (a) 1.95 days (c) 2.8 days (e) 4 days
   (b) 2.1 days   (d) 3.3 days

3. x completes typing of a certain no. of pages in 10 days; B takes half this time. In one day if both are engaged, how much work's completed?

   (a) $\frac{1}{15}$th (b) $\frac{1}{10}$th (c) $\frac{1}{5}$th (d) $\frac{3}{10}$th (e) $\frac{2}{5}$th •

4. If A works for 9hr/day, he can complete a certain work in 10 days. B completes in 15 days working 8 hr/day. If both A and B are engaged working for 10 hr/day. How many days do they take?

   (a) $5\frac{1}{5}$ days (c) $5\frac{1}{14}$ days (e) $6\frac{1}{2}$ days
   (b) $5\frac{1}{7}$ days (d) 6 days

5. A completes 35 pages in 5 hours on a computer. B does 36 pages in 6 hours. Together how much time do they take to complete 390 pages, if they work for 6hr / days?

   (a) 1 day   (b) 2 days   (c) 3 days   (d) 4 days   (e) 5 days

6. P takes thrice as much time as Q and four times as much time as R takes to complete a certain job. Together they finish in 1.5 days. How much time does Q take to complete it alone?

   (a) 1/2 day   (b) 1 day   (c) 2 days   (d) 3.5 days   (e) 4 days

7. P alone can print a report in 15 days. Q alone in 20 days. P, Q, R together can do it in 5 days. How long does R take alone to do this job?

   (a) 9 days   (b) 12 days   (c) 15 days   (d) 18 days   (e) 21 days

8. P can complete a job in 30 day; (Q + R) can complete in 18 days; P and R in 15 days. How long does Q take to complete the job alone?

   (a) 39 days   (b) 40 days   (c) 42 days   (d) 45 days   (e) 48 days

9. P can do a job in 32 days, Q in 24 days; They together work for 8days a similar job. What is the uncompleted portion of the work?

   (a) $\frac{1}{12}$th part   (c) $\frac{1}{4}$th part   (e) $\frac{5}{12}$th part

   (b) $\frac{1}{6}$th part   (d) $\frac{1}{3}$rd part

10. P is 30% more efficient than Q. If Q can complete a job in 39 days. How much time does P take?

    (a) 22 days   (b) 24 days   (c) 26 days   (d) 28 days   (e) 30 days

11. P and Q can do a job in 4/3 days. P is $1\frac{1}{3}$ times more efficient than Q. P alone can do the job in how many days?

(a) 1 day   (b) 1.2 days (c) 1.4 days (d) 1.5 days (e) 1.7 days

12. P can do a work in the same time as (Q and R); (P + Q) can do the same in 12 days. R alone does in 60 days. In how many days can Q do it alone?

    (a) 22 days   (b) 24 days   (c) 26 days   (d) 28 days   (e) 30 days

13. A and B can complete the DTP work on a project report in 4 and 5 days respectively. They are engaged together and are paid Rs.266.67. What are the amounts A and B get?

    (a) 128.15, 138.52   (c) 138.15, 128.52   (e) 148.15, 118.52
    (b) 133.15, 133.52   (d) 143.15, 123.52

14. P and Q can complete a job in 35 and 30 days individually. After together working for a certain number of days, P has to leave and Q alone completes the remaining work in 17 days. Find the number of days they have worked together.

    (a) 4 days   (b) 5 days   (c) 6 days   (d) 7 days   (e) 8 days

15. P, Q, R do a work together and the total payment was Rs.740/-. P and Q together completed 27/37th work, and Q and R completed 25/37th work. What amount does R get?

    (a) Rs.150/- (b) Rs.180/- (c) Rs.200/- (d) Rs.240/- (e) Rs.300/-

Solutions

1. (1 man + 1 woman)'s 1 day's work = 1/5; 1 man's 1 day's work = 1/8 work.

   1 woman alone in one day can do $\frac{1}{5} - \frac{1}{8} = \frac{8-5}{40} = \frac{3}{40}$ work. So she alone will take 40/3 days to complete the job.

2. Work done by each in one day separately respectively is $\frac{1}{28}, \frac{1}{14}, \frac{1}{7}$.
   Together they will do per day $\frac{1}{28} + \frac{1}{14} + \frac{1}{7} = \frac{1+2+4}{28} = \frac{7}{28} = \frac{1}{4}$
   of work x, y, z. So take 4 days to complete together

3. (x + y) in one day $\frac{1}{10} + \frac{1}{5} = \frac{1+2}{10} = \frac{3}{10}$

4. A's one day's work @ $9\frac{hr}{day} = \frac{1}{10}$ th; one hour's work $= \frac{1}{10} \times \frac{1}{9} = \frac{1}{90}$ th

   B's one day's work; @ $8\frac{hr}{day} = \frac{1}{15}$ th, one hours's work $= \frac{1}{15} \times \frac{1}{8} = \frac{1}{120}$ th

   (A + B) in one hour can complete $\frac{1}{90} + \frac{1}{120} = \frac{4+3}{360} = \frac{7}{360}$ th work

   ∴ In 10 days → $\frac{7}{36}$ th per day

   ∴ They together take **36/7 days**

5. A's one hour work → 7 pages; B's → 6 pages

   (A + B) in one hour: 13 pages; in 6 hours → 78 pages;

   Thus 78 pages → 6 hour (or) 1 day

   ∴ 390 pages → $\frac{1}{78} \times 390 =$ **5 days**

6. Let P take 'P' days, Q → $\frac{P}{3}$ days and R → $\frac{P}{4}$ days to complete individually. Given (P + Q + R) together will complete in 3/2 days; P'S one day's work $= \frac{1}{P}$; Q'S one day's work: $\frac{3}{P}$; $R'S \frac{4}{P}$; in one day

   (P + Q + R) complete $\frac{1}{P} + \frac{3}{P} + \frac{4}{P} = \frac{8}{P}$ th. They together complete in $\frac{P}{8}$ days $= \frac{3}{2}$ days (given)

   ∴ $P = 8 \times \frac{3}{2} = 12$ days

   ∴ Q will complete in P/3 = **4 days, alone**

7. P's one days work → $\frac{1}{15} \frac{work}{day}$; $Q's$ → $\frac{1}{20} \frac{work}{day}$ and (P + Q + R)$\frac{1}{5} \frac{work}{day}$

   ∴ R's one day's work $= \frac{1}{5} - \frac{1}{15} - \frac{1}{20} = \frac{12-4-3}{60} = \frac{5}{60}$

   ∴ R takes **12 days** alone

8. P's one day work = $\frac{1}{30}\frac{work}{day}$; $(Q+R) \to \frac{1}{18}\frac{work}{day}$;

   $(P+R) \to \frac{1}{15}\frac{work}{day}$

   ∴ R's one day work = $\frac{1}{15} - \frac{1}{30} = \frac{2-1}{30} = \frac{1}{30}$

   Q's one days work = $\frac{1}{18} - \frac{1}{30} = \frac{5-3}{90} = \frac{2}{90}$

   Q can alone complete in $\frac{90}{2} = $ **45** days

9. P in one day does $\frac{1}{32}$th part; Q in $\frac{1}{24}$th part; (P + Q) in one day
   = $\frac{1}{32} + \frac{1}{24} = \frac{3+4}{96} = \frac{7}{96}$th; in 8 days → $\frac{7}{96} \times 8 = \frac{7th}{12}$

   Remaining part = $\frac{5}{12}$th **work**

10. P takes $\frac{100}{130} \times 39 = $ **30 days** ans

    Check Q takes 3P days. In one day P does 1/30th and Q 1/39th work

    ∴ $\frac{n_P}{n_Q} = \frac{\left(\frac{1}{30}\right)}{\left(\frac{1}{39}\right)} = \frac{39}{30} = 1.3 \, QED$

11. If Q takes n days, P takes $\frac{3}{5}$ n days, P's 1 day work: = $\frac{5}{3n}$; Q's = $\frac{1}{n}$th; (P + Q)'s one day work = $\frac{5}{3n} + \frac{1}{n} = \frac{5+3}{3n} = \frac{8}{3n}$th this is given as $\frac{4}{3}$ days

    ∴ $\frac{8}{3n} = \frac{4}{3}$ days

    ∴ $n = \frac{3 \times 8}{3 \times 4} = 2$ days

    P alone does in $\frac{3}{5} \times 2 = \frac{6}{5}$ **days**

12. Let P do in n days; (Q + R) also in n days P's 1 day's work $= \frac{1}{n}$th; (Q + R) in one day $= \frac{1}{n}$th

∴ (P + Q + R) in one day $= \frac{2}{n}$th; (i) Now (P + Q) in one day; $\frac{1}{12}$th;

R in one day $\frac{1}{60}$th

∴ (P + Q + R) in one day $= \frac{1}{12} + \frac{1}{60} = \frac{5+1}{60} = \frac{6}{60}$

$= \frac{1}{10} = \frac{2}{n}$th from (i)

∴ n = 20 days

∴ P does in 20 days;

(P + Q) in one day $= \frac{1}{12}$th, P in one day $= \frac{1}{20}$th

∴ Q in one day $= \frac{1}{12} - \frac{1}{20} = \frac{5-3}{60} = \frac{2}{60} = \frac{1}{30}$

∴ Q alone **30 days**

13. A's Per day work $= \frac{1}{4}$th; B → $\frac{1}{5}$th per day

(A + B) =→ $\frac{1}{4} + \frac{1}{5} = \frac{5+4}{20} = \frac{9}{20}$th

∴ They together will complete in $\frac{20}{9}$ days. They will share the payment in the ratio of 5 : 4; Total = 266.67 = $\frac{800}{3}$; A's share = $\frac{5}{9}$th and B's = $\frac{4}{9}$th.

They will get $\frac{4000}{27}$ and $\frac{3200}{27}$ ie **148.15** and **Rs.118.52/-** respectively.

Check: Total = 266.67 Rs.QED

14. P in one day $= \frac{1}{35}$th; Q in one day $= \frac{1}{30}$th;

(P + Q) in one day $= \frac{1}{35} + \frac{1}{30} = \frac{6+7}{210} = \frac{13}{210}$th;

If they work for 'n' days together they finish $\left(\frac{13n}{210}\right)$th work;

Remaining work = $1 - \dfrac{13n}{210} = \dfrac{210 - 13n}{210}$

This Q alone completes in 17 days. But, In 17 days Q's work = $\dfrac{17}{30}$th

∴ This should be equal to $\dfrac{210 - 13n}{210}$

∴ $\dfrac{17}{30} = \dfrac{210 - 13n}{210}$

∴ $(7 \times 17) = 210 - 13n \Rightarrow 119 = 210 - 13n \Rightarrow 13n = 91$

∴ $n = $ **7 days**

15. $(P+Q) = \dfrac{27}{37}$th; $Q+R \rightarrow \dfrac{25}{37}$

∴ $(P+2Q+R) \rightarrow \dfrac{52}{37}$th

But $(P + Q + R)$ have done, of course, full i.e $\dfrac{37}{37}$th work

∴ Q's work is $\dfrac{15}{37}$; Hence R's work is $\dfrac{25}{37} - \dfrac{15}{37} = \dfrac{10}{37}$th

R gets $\dfrac{10}{37} \times$ Rs.740/− = Rs.200/− ans

Key

1)	e	2)	e	3)	d	4)	b	5)	e	6)	e
7)	b	8)	d	9)	e	10)	e	11)	b	12)	e
13)	e	14)	d	15)	c						

17

Pipes & Water Tanks

Exercises

1. If two supply pipes fill a tank in 4 and 5 hours respectively and a pipe R can empty it in 8 hours, what is the time to fill if all the 3 are simultaneously opened?

   (a) $2\frac{8}{13}$ hours   (c) 3 hours   (e) $3\frac{5}{13}$ hours

   (b) $2\frac{11}{13}$ hours   (d) $3\frac{1}{13}$ hours

2. A tank can be filled in 5hours by a tap, it can be emptied in 10hr by another. What is the filling time if both are opened at the same time?

   (a) $1\frac{1}{2}$hr hours   (c) 2 hours   (e) 2.6 hours

   (b) $1\frac{4}{5}$ hours   (d) $2\frac{1}{2}$ hours

3. A cistern is half full. A pipe P can fill the cistern in 8 minutes and another pipe Q can empty it is 4 minutes. If both are opened simultaneously, how long does it take to fill or empty the cistern?

   (a) 2 min to full   (c) 3 min to fill   (e) 4 min to empty

   (b) 2 min to empty   (d) $3\frac{1}{2}$ min to empty

4. There are three supply pipes to a cistern. The first two opened simultaneously fill the tank in the same time as 6/5 times the time taken by the third alone. The 2nd pipe alone fills the tank in 4hours faster than the first one; and it fills in 4hours slower than the third alone. In what time does the first pipe alone fill the cistern?

   (a) 4hr   (b) 6hr   (c) 8hr   (d) 10hr   (e) 12hr

5. Two pipes fill a cistern together in $\frac{4}{3}$ hours. Pipes y takes more time than pipe x alone to fill. What time does x take alone to fill the cistern?

   (a) 1 hour (c) 1.75 hours (e) $2\frac{1}{2}$ hours
   (b) 1.5 hours (d) 2 hours

6. A tank is filled in $2\frac{2}{7}$ hours by three pipes x, y, z. Pipe z fills twice as fast as y and y fills twice as fast as x. How much time does x alone take for complete filling?

   (a) 2hours (b) 4hours (c) 8hours (d) 16hours (d) 32hours

7. Pipe x can fill a cistern 4 times faster than y. Together they can fill the cistern in 48min. What is the time of filling by the slower filling pipe alone?

   (a) 1hour (b) 2.2hour (c) 3.3hour (d) 4hour (e) 4.4hour

8. 10 buckets fill a cistern. When the capacity of a bucket is 12l. How many buckets are needed if the capacity of the buckets is 8l?

   (a) 9 buckets (c) 14 buckets (e) 16 buckets
   (b) 12 buckets (d) 15 buckets

9. Three pipes x, y, z can fill a tank in 5 hours. After simultaneously opening then, z is closed after 3 hours. The tank is filled in four more hours. In how much time can z alone fill the tank?

   (a) 7 hours (c) 8 hours (e) 11 hours
   (b) 7 1/2 hours (d) 10 hours

10. Two pipes x and y can fill a cistern in 10min and 15 min. After operating both for 3min y is shut. In what total time will the cistern be full?

    (a) 8min (b) 10min (c) 11min (d) 12min (e) 14min

11. Two pipes x and y can fill a cistern in 50min and 25min respectively. What time does it take to fill the cistern if y is used for half the time and x and y are together used for the other half time?

    (a) 18min (b) 20min (c) 22min (d) 25min (e) 30min

Quantitative Aptitude

12. Two pipes x and y can fill a tank in 40 and 50min respectively. Both are opened and after some time x is put off. The total time of filling the tank was 30min. After how much time was x put off?

    (a) 10min (b) 14min (c) 16min (d) 18min (e) 20min

13. Two pipes x and y can fill a tank in 8hours and 12hours. An outlet pipe z can empty the tank in 16hours. All the three operate simultaneously for 6hours and then z is closed. What is the time in which the tank can be full?

    (a) $6\frac{2}{5}$hr (b) $6\frac{3}{5}$hr (c) $6\frac{4}{5}$hr (d) 7hr (e) $7\frac{1}{5}$hr

14. Pipes x and y can fill a tank in 8 and 12hrs respectively. They are worked on alternately, starting with x. In how long a time the tank will be full?

    (a) 8 hours   (c) $9\frac{1}{4}$ hours   (e) 10 hours
    (b) $8\frac{3}{4}$ hours   (d) $9\frac{1}{2}$ hours

15. Capacity of bucket B_1 is twice that of B_2. If 75 buckets type B_1 fill a drum, how many times ($B_1 + B_2$) fill it?

    (a) 20 (b) 25 (c) 30 (d) 40 (e) 50

Solutions

1. $\frac{1}{r} = \frac{1}{4}$th per hour; $\frac{1}{q} = \frac{1}{5}$th per hour; $\frac{1}{r} = \frac{-1}{8}$th per hour.

   Hence the required time can be found as the inverse of $\left(\frac{1}{p} + \frac{1}{q} - \frac{1}{r}\right)$

   i.e., of $\left(\frac{1}{4} + \frac{1}{5} - \frac{1}{8}\right) = \frac{10 + 8 - 5}{40} = \frac{13}{40}$

   ∴ Time required is $\frac{40}{13}$ hours = $3\frac{1}{13}$ hours

2. $\dfrac{1}{\left(\dfrac{1}{5} - \dfrac{1}{10}\right)} = \dfrac{1}{\left(\dfrac{10-5}{10}\right)} = \dfrac{1}{\left(\dfrac{5}{10}\right)} = \dfrac{10}{5} = 2\text{hours}$

3. In one minute $\dfrac{1}{8} - \dfrac{1}{4} = \dfrac{1-2}{8} = \dfrac{-1}{8}$ is the part emptied ($\because$ it is negative)

$\therefore$ It takes $\dfrac{1}{\left(\dfrac{1}{8}\right)} = 4$ minutes to empty the cistern

4. Let t_1, t_2, t_3 be the times of filling for the three individually. Given $t_1 - 4 = t_2$; and $t_2 = t_3 + 4$.

Also in 1 hour first two will fill $\left(\dfrac{1}{t_1} + \dfrac{1}{t_2}\right)^{th}$ part $= \left(\dfrac{1}{t_2+4} + \dfrac{1}{t_2}\right) = \left[\dfrac{2t_2+4}{t_2(t_2+4)}\right]$ th part i.e they take $\left[\dfrac{t_2(t_2+4)}{2t_2+4}\right]$ hours

This is given as $1.2\, t_3 = 1.2\,(t_2 - 4)$

$\therefore \dfrac{t_2(t_2+4)}{2t_2+4} = \dfrac{6}{5}(t_2-4)$

$\therefore 5t_2(t_2+4) = 12(t_2+2)(t_2-4) \Rightarrow 5t_2^2 + 20t_2 = 12t_2^2 - 24t_2 - 96$

$\Rightarrow 7t_2^2 - 44t_2 - 96 = 0$
$\Rightarrow 7t_2^2 - 56t_2 + 12t_2 - 96 = 0$
$\Rightarrow 7t_2(t_2-8) + 12(t_2-8) = 0$
$\Rightarrow (7t_2+12)(t_2-8) = 0$

$t_2 = \left(-\dfrac{12}{7}\right)$ is not admissible

```
7 × 96
14 × 48
28 × 24
42 × 16
56 × 12
```

$\therefore t_2 = 8$ hours; then $t_1 = 12$ hours; $t_3 = 4$ hours

5. Let x and y hours be the individual times to fill. Then $\dfrac{1}{x}$ and $\dfrac{1}{y}$ will be the parts filled separately per hour

$\therefore \dfrac{1}{x} + \dfrac{1}{y} = \dfrac{x+y}{xy}$ is the part filled in 1 hour together. It is given that $[xy \div (x+y)]$ hours is the time for fully filing that $\dfrac{xy}{x+y} = \dfrac{4}{3}$ hours; also $y = x + 2$

$\therefore \dfrac{x(x+2)}{2x+2} = \dfrac{4}{3}$

$\therefore 3x^2 + 6x = 8x + 8$

$\Rightarrow 3x^2 - 2x - 8 = 0$

$3x^2 - 6x + 4x - 8 = 0$ $\quad\quad\quad\quad\quad\quad\quad\quad \begin{array}{|l} 3 \times 8 \\ 6 \times 4 \end{array}$

$\therefore 3x(x-2) + 4(x-2) = 0$

$\Rightarrow (3x+4)(x-2) = 0$

$x = -\dfrac{4}{3}$ is not admissible

$\therefore x = 2$ hours and $y = 4$ hours

6. Let x, y, z be the hours to fill by x, y, z;

Given $z = \dfrac{y}{2}$; $y = \dfrac{x}{2}$ and $\dfrac{1}{x} + \dfrac{1}{y} + \dfrac{1}{z} = \dfrac{1}{4z} + \dfrac{1}{2z} + \dfrac{1}{z}$

$\dfrac{1+2+4}{4z} =$ (since $x = 2y$; $y = 2z$, $x = 4z$) $\dfrac{7}{4z}$ th part in one hour by all the three.

All the 3 take $\dfrac{4z}{7}$ hours given as $\dfrac{16}{7}$ hours

$\therefore z = 4$ hours $\quad y = 8$ hours, $x = 16$ hours

7. Let x, y be the time in minutes to fill by x and y separately. In one minute the part filled in is $\dfrac{1}{x} + \dfrac{1}{y}$; it is given $x = \dfrac{y}{4} \Rightarrow y = 4x$

$\therefore$ In one minute $\dfrac{1}{x} + \dfrac{1}{4x} = \dfrac{4+1}{4x} = \dfrac{5}{4x}$

$\therefore$ It takes $\dfrac{4x}{5}$ minutes to fill by both simultaneously operating;

This = 48 min

$\therefore x = \dfrac{5}{4} \times 48 = 60$ min = 1 hours

y = 4 hours

8. Lesser bucket capacity → more buckets IP $8l : 12l = 10 : (6)$ capacity and no of buckets respectively (Left, right items).

Pipes & Water Tanks

$\therefore$ $8b = 10 \times 12$

$\therefore$ $b = \dfrac{10 \times 12}{8} = 10 \times \dfrac{3}{2} = 15$ buckets

9. Let x, y, z be the hours in which they individually fill. Then in one hour all the three fill $\left(\dfrac{1}{x} + \dfrac{1}{y} + \dfrac{1}{z}\right)^{th}$ part.

In 3hours they fill $3 \times \left[\dfrac{1}{x} + \dfrac{1}{y} + \dfrac{1}{z}\right]^{th}$ part.

Since the three can fill in 5hours $\left[\dfrac{1}{x} + \dfrac{1}{y} + \dfrac{1}{3}\right]^{-1} = 5$ hours.

$\therefore 3\left[\dfrac{1}{x} + \dfrac{1}{y} + \dfrac{1}{z}\right] = \dfrac{3}{5}$th. Remaining $\dfrac{2}{5}$th part is filled by x and y in 4hours. In one hour these two fill $\left(\dfrac{1}{x} + \dfrac{1}{y}\right)^{th}$ part.

This is then $\dfrac{1}{4} \dfrac{2}{5} = \dfrac{1}{10}$ th part

$\therefore \dfrac{1}{z} = \left(\dfrac{1}{x} + \dfrac{1}{y} + \dfrac{1}{z}\right) - \left(\dfrac{1}{x} + \dfrac{1}{y}\right) = \dfrac{1}{5} - \dfrac{1}{10} = \dfrac{2-1}{10} = \dfrac{1}{10}$

z = 10hours

10. In one minute x alone fills in one minute;

Both fill $\dfrac{1}{10} + \dfrac{1}{15} = \dfrac{3+2}{30} = \dfrac{5}{30} = \dfrac{1}{6}$th part; $\dfrac{1}{x} = \dfrac{1}{10}^{th}$ part.

In 3min, $\dfrac{1}{2}$ is full. Remaining is $\dfrac{1}{2}$.

This is filled in $\left[\dfrac{\frac{1}{2}}{\frac{1}{10}}\right] = 5$min;

Total time = 3+5 = 8min

Check $\left[\left(8 \times \dfrac{1}{10}\right) + \left(3 \times \dfrac{1}{15}\right)\right] = \dfrac{4}{5} + \dfrac{1}{5} =$ full

11. Let x and y minutes (50 and 25min) be the times to fill individually in one minute the rate of filling is $\frac{1}{50}$ th part and $\frac{1}{25}$ th part. Let it take 't' minutes to fill the cistern as is mentioned. For $\frac{t}{2}$ minutes only y is functioning and for the other, $\frac{t}{2}$ minutes both x and y are functioning in $\frac{t}{2}$ minutes $\left(\frac{1}{25} \cdot \frac{t}{2}\right)$ th part is filled by y alone and in $\frac{t}{2}$ minutes $\left(\frac{3 \text{ part}}{50 \text{ min}} \cdot \frac{t}{2} \text{ minutes}\right)$ is filled by x & y

$$\left[\text{since } \left(\frac{1}{50} + \frac{1}{25}\right) = \frac{1+2}{50} = \frac{3}{50} th\right) \frac{\text{part}}{\text{min}}\right]$$

Sum $= \frac{t}{50} + \frac{3t}{100} = \frac{2t+3t}{100} = \frac{5t}{100}$

Given sum = full cistern (1)

Thus $\frac{t}{50} + \frac{3t}{100} = 1 \Rightarrow \frac{2t+3t}{100} = 1$

$\therefore$ 5t = 100

$\therefore$ t = 20min

Check In 10min y fills $\frac{10}{25}$ th and in 10min (x + y) fill $\left(\frac{3}{50} th \frac{\text{part}}{\text{min}}\right) \times$ 10 min = $\frac{30}{50}$;

Total $= \frac{20}{50} + \frac{30}{50} =$ full

12. Let x and y in minutes be the times to fill individually.

In one minute together $\frac{1}{x} + \frac{1}{y}$ i.e $\frac{1}{40} + \frac{1}{50} = \frac{5+4}{200} = \frac{9}{200}$ th part is filled;

In t min. they together fill $\frac{t \times 9}{200}$ th part and if y works alone for (30 - t) mits

$\therefore \frac{9t}{200} + \frac{(30-t)}{50} = 1 \Rightarrow 9t + 120 - 4t = 200$

$\therefore$ 5t = 80

Pipes & Water Tanks 131

$\therefore t = 16$ min

Check $\left(\dfrac{9}{200} \times 16\right) + \dfrac{14}{50} = \dfrac{144 + 56}{200} = 1$ QED

13. In one hour the net rate of filling is $\left[\dfrac{1}{8} + \dfrac{1}{12} - \dfrac{1}{16}\right]$; in 6 hours the part filled is $\left(\dfrac{1}{8} + \dfrac{1}{12} - \dfrac{1}{16}\right) 6$; in say t more hours the filling part is $\left[\dfrac{1}{8} + \dfrac{1}{12}\right] t$ now the tank is full. The total time is $(6 + t)$ hours

$\therefore \left(\dfrac{1}{8} + \dfrac{1}{12} - \dfrac{1}{16}\right) 6 + \left(\dfrac{1}{8} + \dfrac{1}{12}\right) t = 1 \Rightarrow \dfrac{3}{4} + \dfrac{1}{2} - \dfrac{3}{8} + \dfrac{t}{8} + \dfrac{t}{12} = 1$

$\therefore \dfrac{t}{8} + \dfrac{t}{12} = 1 - \dfrac{3}{4} - \dfrac{1}{2} + \dfrac{3}{8} = \dfrac{8 - 6 - 4 + 3}{8} = \dfrac{1}{8}$

$\Rightarrow \dfrac{3t + 2t}{24} = \dfrac{5t}{24} = \dfrac{1}{8} = \dfrac{3}{24}$

$\therefore t = \dfrac{3}{5}$ hrs

$\therefore$ Required is $6\dfrac{3}{5}$ hours

14. x and y in two hours fill $\dfrac{1}{8} + \dfrac{1}{12} = \dfrac{3+2}{24} = \dfrac{5}{24}$ th part.

In 4 sets of two hours they fill $4 \times \dfrac{5}{24} = \dfrac{5}{6}$ th part remaining $= \dfrac{1}{6}$ th part;

In one hr x fills $\dfrac{1}{8}$ th still remaining $= \dfrac{1}{6} - \dfrac{1}{8} = \dfrac{4-3}{24} = \dfrac{1}{24}$.

This is filled by y in $\left[\dfrac{\frac{1}{24}}{\frac{1}{12}}\right] = \dfrac{1}{2}$ hour.

Total time is $\dfrac{xy}{2hr}, \dfrac{xy}{2hr}, \dfrac{xy}{2hr}, \dfrac{xy}{2hr}, \dfrac{xy}{1hr}, \dfrac{1}{2}$

Total = 9.5 hrs

∴ In 10 hours x fills = $\frac{10}{8}$; In 9 hr y fills = $\frac{9}{12} = \frac{3}{4}$

Duration → 2hu 2hu 1hour

x → 10hours; y → 9hours

Hence total = 9.5 hours
One tank is full
Total = 9.5 hours

Check → $\left(5 \times \frac{1}{8}\right) + \left(\frac{9}{2} \times \frac{1}{12}\right) = \frac{5}{8} + \frac{3}{8} = 1$

$\underset{x}{\downarrow}\, \underset{y}{5 + 4\frac{1}{2}} = 9\frac{1}{2}$ hours

15. Let capacities be B_1 and B_2 units

$B_1 = 2 B_2; B_2 = \frac{B_1}{2}$

∴ $(B_1 + B_2) = \left(B_1 + \frac{B_1}{3}\right) = \frac{3B_1}{2}$ units

$\underset{\text{unit}}{U} = \frac{3B_1}{2}$ or $B_1 = \frac{2U}{3}$

Full drum = 75 B_1 units or $\left[\frac{\text{full drum}}{B_1}\right] = 75$; (or)

$\frac{\text{full drum}}{2U} \times 3 = 75$

∴ full drum = $75 \times \frac{2U}{3} = 50U$

⇒ $50 (B_1 + B_2) = 50$ times is the answer

Key

1)	d	2)	c	3)	e	4)	c	5)	d	6)	d
7)	d	8)	d	9)	d	10)	a	11)	b	12)	c
13)	b	14)	d	15)	e								

18

Time & Distance

Exercises

1. A cyclist covers a distance of 640 m in 2 min 40 sec. What is the speed of the cyclist in kmph?

   (a) $13\dfrac{km}{hr}$ (b) $13\dfrac{1}{5}\dfrac{km}{hr}$ (c) $14\dfrac{km}{hr}$ (d) $14\dfrac{2}{5}\dfrac{km}{hr}$ (e) $14\dfrac{4}{5}\dfrac{km}{hr}$

2. A sprinter runs at the speed of 24 km./hr. and walks back at the speed of 4 km./hr. What total time did he take to go to and for, a total distance of 1200 m?

   (a) $9\dfrac{3}{4}$ min   (c) 10.5 min   (e) 12.80 min
   (b) 10 min   (d) 11.25 min

3. If a person walks @ 4 km./hr. he misses a train, by 10 minutes. If he walks @ 6 km./hr, he reaches 5 minutes before hand. Find the distance to the station

   (a) 1/2 km   (b) 1 km   (c) 2 km   (d) 3 km   (e) 3.75 km

4. Going with a speed equal to 4/5th of usual speed, a car reaches from the officers house to the office late by 8 minutes. What is the usual time it takes to cover the distance?

   (a) 27 min   (b) 29 min   (c) 31 min   (d) 32 min   (e) 36 min

5. A car covered a certain distance in 19 hours. Its speed is 40 km per hour during the first 3/4 of the distance and 50 km per hour during the remaining portion. What is the distance covered?

(a) 750 km (b) 800 km (c) 850 km (d) 950 km (e) 1000 km

6. The time taken by a person to go by bus on one side and return by cycling on the return side is 16/5 hours. If he goes both sides by bus he would have taken 12/5 hours. If he cycles both ways, what time does he take?

(a) 1 hr (b) 2 hr (c) 3 hr (d) 4 hr (e) $5\,{}^3/_4$ hr

7. A passenger train started at a station P moving towards north. An express train starts at P towards north 2 hours After 4hours. Of travel E crosses P. The express train moves at a speed of 90kmph. What is the speed of the passenger train?

(a) 50 kmph (b) 60 kmph (c) 75 kmph (d) 90 kmph

8. A race car starts with a speed of 80 kmph and increases its speed at the end of every hour by 10 kmph. In how many hours does it cover 560 km?

(a) 4.9 hr (b) 5.1 hr (c) 5.2 hr (d) 5.5 hr (e) 5.9 hr

9. What time does a boy take to complete a run around a circular path with radius 7 m; If his speed is 8 km/h ?

(a) 19.6 sec (b) 19.8 sec (c) 21.2 sec (d) 23.5 sec (e) 24.6 sec

10. Sound travels with a level of 1100 ft per sec in air. If a person sees a 'wood cutter' striking a tree from a distance and hears the sound of the axe cutting the three after 2 seconds, what is the viewer's distance from the wood cutter?

(a) 2200 ft (b) 2225 ft (c) 2295 ft (d) 2410 ft (e) 2450 ft

Time & Distance   135

11. Two race cars A, B completed the following distances (1) $35\dfrac{m}{se}$ (2) 375 km in 3 hours respectively. Which has won?

   (a) A
   (b) B
   (c) both equal speeds
   (d) data in complete

12. A person in a moving train counts 31 electric poles in one minute. If they are at an equal distance of 60 m apart, what is the speed of the train?

   (a) 100 kmph
   (b) 102 kmph
   (c) 104 kmph
   (d) 106 kmph
   (e) 108 kmph

13. Car A covers 900 meters per minutes on an average and car B covers 45.5 km per 50 minutes. Which is going with a faster speed?

   (a) A   (b) B   (c) equally fast   (d) can't conclude

14. A person jogging early morning crosses a 1350 m long street in 9 min. What is his speed in kmph?

   (a) 4 kmph
   (b) 6 kmph
   (c) 8 kmph
   (d) 9 kmph
   (e) 12 kmph

15. A car moving at a speed of 99 kmph can reach the destination at 550 m in what time?

   (a) 15 min   (b) 20 min   (c) 25 min   (d) 30 min   (e) 37 min

Solutions

1. Speed of the cycle = $\dfrac{640m}{160 \sec} = \dfrac{4m}{\sec} = 4m \times 3600 \dfrac{m}{hours}$

   $\Rightarrow 4 * \dfrac{3600}{1000}$ in (kmph); $= 4 * \dfrac{36}{10} = 4 * \dfrac{18 km}{5 hr} = \dfrac{72}{5} = 14\dfrac{2}{5}$ kmph

   Note: $\left[\dfrac{1m}{\sec} = \dfrac{18}{5}\dfrac{km}{hour}\right]$

2. $\dfrac{1 \text{ km}}{\text{hour}} = \dfrac{1000m}{3600 \sec} = \dfrac{5}{18} \dfrac{m}{s}$ time he runs

$\dfrac{600m}{5m} \times \dfrac{18}{24} \sec = \dfrac{120 \times 18}{60 \times 24} = \dfrac{36 \min}{24} = \dfrac{3}{2} \min$

Time he walks back = $\dfrac{600m \times 18}{4 \times 5m} = \dfrac{600 \times 18}{4 \times 5 \times 60} = 9 \min$

Total time = $10\dfrac{1}{2} \min$

3. Let x km be the distance $\dfrac{x}{4}$ ho $- \dfrac{x}{6}$ ho $= \left(\dfrac{x}{4} - \dfrac{x}{6}\right)$ 60 min = 10+5 = 15 min

∴ $\dfrac{3x - 2x}{12} = \dfrac{15}{60} hr = \dfrac{1}{4} hr \Rightarrow \dfrac{x}{12} = \dfrac{1}{4}$

∴ $x = \dfrac{12}{4} = 3$ km

4. Speed = $\dfrac{\text{distance}}{\text{time}}$, less speed → more time (IP: Inverse Proportion);

Present speed = 4/5 usual speed

∴ Present coverage time = (5/4) × usual coverage time; (present coverage time - usual time = 8 min

∴ (5/4 usual coverage time) - usual time = 8 min

⇒ 1/4 usual time = 8 min

∴ Usual time = 32 min

5. Let x km be the distance:

$\left(\dfrac{3}{4}x\right)\dfrac{1}{40}hr + \left(\dfrac{1x}{4}\right)\dfrac{1}{50} = \dfrac{3x}{160} + \dfrac{x}{200} = \dfrac{15x + 4x}{800} = \dfrac{19x}{800} = 19 \text{hours}_{\text{given}}$

∴ x = 800 km

6. Let the speed of bus be b - $\dfrac{km}{hr}$, speed of cycle c $\dfrac{km}{hr}$,

Let x be the distance in km one way

∴ $\dfrac{x}{b} + \dfrac{x}{c} = \dfrac{16}{5}$; also $\dfrac{2x}{b} = \dfrac{12hr}{5}$

$$\therefore \frac{x}{b} = \frac{6}{5}$$

$$\therefore \frac{x}{c} = \frac{16}{5} - \frac{6}{5} = \frac{10}{5};$$

$$\therefore \frac{2x}{c} = \frac{20}{5} = 4 \text{ hours}$$

7. If they cross at Q, E has moved for 4 hour P travelled for 6 hours with a speed of s kmph say

$$\therefore 6S = 4 \times 90 \text{ km}$$

$$\therefore S = \frac{4 \times 90}{6}$$

$$S = 60 \text{ kmph}$$

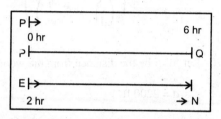

8. The speeds are in a arithmetic progression first term a = 80 kmph; common difference d = 10 kmph

Let us first find the time up to 500 kmph being a round a figure smaller than 560 km

Term in the A.P = a + (n - 1) d

Sum to n terms = $\frac{n}{2}$ [2a + (n - 1)d] = 500 km

$\Rightarrow \frac{n}{2}$ [160 + (n - 1)10] = 500

$\Rightarrow$ 80n + 5n (n - 1) = 500

$\Rightarrow 5n^2 + 75n - 500 = 0$

$\Rightarrow n^2 + 15n - 100 = 0$

$\Rightarrow n^2 + 20n - 5n - 100 = 0$

$\Rightarrow$ n (n + 20) - 5 (n + 20) = 0

$\Rightarrow$ (n - 5) (n + 20) = 0

$\Rightarrow \therefore$ n = 5

The speeds are 80, 90, 100, 110, 120. Before the next hour, then the speed becomes 130 kmph, the train covers 560 km since 80 + 90 + 100 + 110 + 120 = 500 km. The 60 km are covered in 1/2 hr

138   Quantitative Aptitude

∴ 5 1/2 hours is the time to cover 560 km ans 5.5 hours

9. Perimeter = $2\pi r = \dfrac{44}{7} \times 7h = 44m$

   Speed = $8\dfrac{km}{hr} = 8 \times \dfrac{5}{18}$ m/sec

   Time taken = $\left(\dfrac{44m}{\dfrac{8}{18} \times 5 \dfrac{m}{sec}}\right) = \dfrac{44 \times 18}{8 \times 5} = \dfrac{99}{5} = 19.8$ sec

10. Let 'd' ft be the distance from the wood cutter $\dfrac{d}{2} \dfrac{ft}{sec} = 1100 \dfrac{ft}{sec}$

    ∴ d = 2200 ft

11. First one = $35 \dfrac{m}{sec}$, second one = $\dfrac{375 km}{3 hr} = 125 \dfrac{km}{hr}$

    First one $\left(in \dfrac{km}{hr}\right) = 35 \times \dfrac{18}{5}$ km/hr = $126 \dfrac{km}{hr} > 125 \dfrac{km}{hr}$

12. Between 4 poles there are three equal distances 'd' each

    ∴ Between 31 poles there will be 30d = 30 × 60 = 1800 m

    ∴ The speed will be 1800 m/minute = $1800 \times 60 \dfrac{m}{hr}$

    $= \dfrac{1800 \times 60}{1000} \dfrac{km}{hr} = 108 \dfrac{km}{hr}$

13. Car A: $900 \dfrac{m}{min} = \dfrac{900}{60}$ meter per sec = $15 \dfrac{m}{sec} = 15 \times \dfrac{18}{5} = 54$ kmph

    Car B: $\dfrac{45.5 \, km}{50 \, minutes} = \dfrac{45.5}{50} \times 60 \dfrac{km}{hour} = 9.1 \times 6$ kmph = 54.6 kmph;

    Car B goes faster

14. $\dfrac{1350 \, m}{9 \, min} = \dfrac{1350 m}{9 \times 60 \, sec} = \dfrac{1350}{9 \times 160} \times \dfrac{18}{5} \dfrac{km}{hour} = \dfrac{9 km}{hour}$

15. Speed $99 \dfrac{km}{hour} = 99 \times \dfrac{5}{18} \dfrac{m}{sec}$; Distance to be covered = 550 m

Time taken = $\dfrac{550 m}{99 \times 5 m} \times 18 = 20$ minutes

Key

1)	d	2)	c	3)	d	4)	d	5)	b	6)	d
7)	b	8)	d	9)	b	10)	a	11)	a	12)	e
13)	b	14)	d	15)	b						

19

Moving Trains

Exercises

1. Two trains of length 270m and 180m moving in the same direction with speeds 81 kmph and 63 kmph. What time the first train takes to cross the second?

    (a) 30 sec (b) 50 sec (c) 60 sec (d) 75 sec (e) 90 sec

2. A train of length 180m is running at 120 kmph. How long does it takes to cross a railway platform 180m long?

    (a) 8.2 sec (b) 9.4 sec (c) 10.8 sec (d) 11.5 sec (e) 12.2 sec

3. A train of length 180m is running at 120 kmph. What is the time to pass a signal post near the track?

    (a) 4.5 sec (b) 5.4 sec (c) 6.3 sec (d) 7.2 sec (e) 8.1 sec

4. A train of length 180m is running at 120 kmph. A person is running at 8 kmph in the opposite direction. In how much time does it cross the person?

    (a) 5 sec (b) 5.06 sec (c) 5.25 sec (d) 5.49 sec (e) 6.25 sec

5. A person is standing on a railway platform 240m long. A train crossing the platform crosses him in how much time? And the platform in how much time?

    (a) 5, 11 sec (c) 5.4, 12.6 sec (e) 5. 8, 14 sec
    (b) 5.2, 12.4 sec (d) 5.6, 13 sec

6. A train running at 62 km/hr takes 10 sec to pass a person running at 8 km/hr. in the same direction. It takes 18 secs to pass a platform. What are the lengths of the train and the platform?

   (a) 140, 170 m   (c) 160, 180 m   (e) 170, 185 m
   (b) 150, 160 m   (d) 165, 175 m

7. A train is moving at 100 km/hr. A person P in it watches a goods train coming in the opposite direction cross him in 5 sec. length of the goods train is 250m. What is the speed of the goods train?

   (a) 50 km/hrs   (c) 60 km/hrs   (e) 80 km/hrs
   (b) 55 km/hrs   (d) 65 km/hrs

8. A train passes a platform 32 sec and a person on the platform in 18 sec. What is the length of the platform? The speed of the train is 72 km/hr.

   (a) 280 m   (b) 300 m   (c) 320 m   (d) 330 m   (e) 335 m

9. A 250m long train crosses a platform in 28s; and crosses a telephone pole in 15s. What is the length of the platform?

   (a) 195 m   (b) 200.4 m   (c) 210.7 m   (d) 216.6 m   (e) 220 m

10. Speed of a train is 72 km/hr. If it takes 8s to pass an electric pole, what is the length of the train?

    (a) 160 m   (b) 165 m   (c) 175 m   (d) 180 m   (e) 190 m

11. A train 250m long running @ 72 km/hr will passes person in?

    (a) 12.5 sec   (c) 15.75 sec   (e) 17 sec
    (b) 14 sec   (d) 16.25 sec

12. A train of length 120m takes 36s to cross a tunnel 330m long. What is its speed?

(a) 35 km/hr. (c) 45 km/hr. (e) 55 km/hr.
(b) 40 km/hr. (d) 50 km/hr.

13. Two trains 150m and 250m long; moving in opposite directions, cross each other in 9sec. Their speeds are in the ratio of 3:2. What is the speed of the faster train?

(a) 65 km/hr. (c) 84 km/hr. (e) 96 km/hr.
(b) 72 km/hr. (d) 90 km/hr.

14. Two trains running in opposite directions cross a person on the platform in 31s and 17 respectively. They cross each other in 25 sec. The ratio of their speeds is

(a) 2:1   (b) 3:2   (c) 4:3   (d) 2:3   (e) 1:3

15. Two trains are running at 70 km/hr. and 50 km/hr. in the same direction. The first one crosses a person sitting in the second train in 18 sec. What is the length of the first train?

(a) 95m   (b) 100m   (c) 120m   (d) 125m   (e) 130m

Solutions

1. Relative speed = 81 − 63 = 18 kmph = $18 \times \dfrac{5m}{18\,sec}$

$\left[\text{Since } 1\dfrac{km}{hr} = \dfrac{1000m}{3600\,sec} \right] = 5\dfrac{m}{sec}$ Time taken to cross

$= \dfrac{(270 + 180)\,m}{5\,\dfrac{m}{sec}}$

$= \dfrac{450}{5} = 90\,sec$

2. Speed of train = 120 km/hr. = $120 \times \frac{5}{18} \frac{m}{sec} = \frac{100}{3} \frac{m}{sec}$

Distance covered in passing over the platform
= 180 + 180 = 360 m

Time to cross = $\frac{360m}{\left(\frac{100}{3}\right)\frac{m}{sec}} = \frac{360 \times 3}{100} = \frac{1080}{100} = 10.8$ sec

```
         |← 180 m →|
  |← 180 m →| p.f.
          T        |←--------→|
                   G           E
```

3. Time taken to pass the signal post is the same as the time taken to cover the length of the train i.e 180 m. speed of the train

= 120 km/hr. = $120 \times \frac{5}{18} \frac{m}{sec} = \frac{100}{3} \frac{m}{sec}$

Time to cross the post = $\frac{180m}{\left(\frac{100}{3}\right)\frac{m}{sec}} = \frac{540}{100} = 5.4$ sec ans

4. Relative speed = 120 + 8 = 128 kmph = $128 \frac{5}{18} \frac{m}{sec}$

Time to cross = $\frac{180m}{128 \times (5m)} \times (18 \, sec) = \frac{36 \times 18}{128} = 5.06$ sec

```
           8 kmph
           👤→
  120 kmph  ← T    180 m
```

5. The train's speed = 120 km/hr. = $120 \times \frac{5}{18} = \frac{600}{18} = \frac{100}{3} \frac{m}{sec}$

It crosses the person in a time equal to $\left[\frac{180m}{\left(\frac{100}{3}\right)\frac{m}{sec}}\right] = \frac{540}{100} = 5.4$ sec

It crosses the platform in a time equal to

$$\frac{[180+240]m}{\left(\frac{100}{3}\right)\frac{m}{\sec}} = \frac{420}{100} \times 3 = \frac{1260}{100} = 12.6 \sec \quad \text{ans}$$

6. Relative speed of the train when the person is running in the same direction $= 62 - 8 = 54\frac{km}{hrs} = 54 \times \frac{5}{18} = 15\frac{m}{\sec}$

In passing the person the train has to cover its own length say 'x' with the relative speed

$$\therefore \frac{xm}{15\frac{m}{\sec}} = \text{time to cross} = 10\sec$$

$$\therefore x = 15 \times 10 = \mathbf{150m} \text{ ans}$$

If y is the length of the platform in m, in crossing the platform the train has to move (x + y) m with its speed i.e 62 km/hr

$$\frac{(150+y)m}{\left(62 \times \frac{5}{18}\right)\frac{m}{\sec}} = 18\sec$$

$$\Rightarrow (150+y) = 18 \times 62 \times \frac{5}{18} m = 310m$$

$\therefore y = \mathbf{160m}$ ans

7. Relative speed of the two trains moving in opposite directions u + v where u - speed of the passenger train, v speed of goods train

Now, $(u+v) = \dfrac{\text{Length of goods train passing by}}{\text{Time to cross}} = \dfrac{250}{5} = \dfrac{m}{\sec} = 50\dfrac{m}{\sec}$

$50 \times \dfrac{18}{5} \dfrac{km}{hrs} = 180 \dfrac{km}{hrs}$; u = speed of passenger train

$= 100\dfrac{km}{hrs}$; Relative speed $(u+v) = 180\dfrac{km}{hrs}$

$\therefore$ v, speed of goods train $= 180 - 100 = 80\dfrac{km}{hrs}$

8. Let l_1 and l_2 m be the lengths of the train and the platform.

Speed of the train $= 72\dfrac{km}{hrs} = 72 \times \dfrac{5}{18} = 20\dfrac{m}{s}$

To cross the person the train has to travel its own length l_1.

$\Rightarrow \left[\dfrac{l_1\ m}{(20\ \frac{m}{sec})}\right] = 18s$

$\therefore l_1 = 360m$; also to cross the pf.

It has to travel $(l_1 + l_2)$ m : $\dfrac{(l_1 + l_2)\ m}{20\ \frac{m}{s}} = 32s$

$\therefore (l_1 + l_2) = 640 \Rightarrow l_2 = 640 - 360 = 280m$

9. $\dfrac{250m}{s\left(\frac{m}{s}\right)} = 15s$; $\dfrac{250 + l}{s\frac{m}{s}} = 28s$

$\therefore 15s = 250$;

$28s = 250 + l$ $\quad \therefore 13s = l$ and $s = \dfrac{250}{15} = \dfrac{50}{3}\ \dfrac{m}{s}$

$\therefore l = 13 \times \dfrac{50}{3} = \dfrac{650}{3}\ m = 216.6m$

10. $s = 72\dfrac{km}{hrs} = 72 \times \dfrac{5}{18} = 20\dfrac{m}{s}$; $\dfrac{l\,m}{20\frac{m}{s}} = 8\,\sec$

$\therefore l = 20 \times 8 =$ **160m ans**

11. $s = 72 \times \dfrac{5}{18} = 20\dfrac{m}{s}$; $\dfrac{250m}{20\frac{m}{s}} = 12\dfrac{1}{2}\,\sec$

12. $\dfrac{(120 + 330)\,m}{s\frac{m}{s}} = 36s$

$\therefore s = \dfrac{450}{36}\ 25\dfrac{m}{s} = \dfrac{25}{2} \times \dfrac{18}{5} = 45\dfrac{km}{hrs}$

13. Ratio and relative speeds 3:2 $\left(\dfrac{m}{s}\right)$, $(3s+2s)$;

Time to cross $= \dfrac{(250 + 150)\,m}{5s}\dfrac{m}{s} = 9s$ (given)

146   Quantitative Aptitude

$$\therefore \frac{400}{5s} = 9 \Rightarrow S = \frac{400}{9 \times 5}. \text{ Speed of the faster train}$$

$$= \frac{400}{9 \times 5} \times 3 = \frac{80}{3} \frac{m}{s} = \frac{18}{3} \times \frac{18}{5} = 96 \frac{km}{hr}$$

14. Let $l_1 m, l_2 m$ be the lengths of the trains and $s_1 \frac{m}{s}$ and $s_2 \frac{m}{s}$ their speeds and let

$$\frac{s_1}{s_2} = r \text{ then } \frac{l_1}{s_1} = \frac{l_1}{s_2 r} = 31s; \frac{l_2}{s_2} = 17s;$$

Then $l_1 = 31 s_2 r$; & $l_2 = 17 s_2$ and $\frac{l_1 + l_2}{s_1 + s_2} = \frac{l_1 + l_2}{s_2(r + 1)} = 25$

$$\therefore \frac{31 s_2 r + 17 s_2}{s_2 (r + 1)} = \frac{s_2 (31 r + 17)}{s_2 (r+1)} = 25$$

$$\Rightarrow 31r + 17 = 25r + 25$$

$$\therefore 6r = 8 \therefore r = \frac{4}{3}$$

15. Let l be its length in m. Then $\dfrac{lm}{(70 - 50) \times \dfrac{5}{18} \dfrac{m}{s}} = 18s \text{ sec}$

The relative speed is $(70 - 50) = 20 \dfrac{km}{hr}$

$$\therefore l = \frac{18 \times 20 \times 5}{18} = 100m \text{ ans}$$

Key

1)	e	2)	c	3)	b	4)	b	5)	c
6)	b	7)	e	8)	a	9)	d	10)	a
11)	a	12)	c	12)	e	14)	c	15)	b

20

Stream Crossing

Exercises

1. A person can ply his boat in still water @ 8 kmph. If the river flows @ 2 kmph; and if he takes 4 hours to go from a place x to another place y and back, what is the distance of xy?

    (a) 4 km   (b) 8 km   (c) 12 km   (d) 15 km   (e) 18 km

2. A boat can travel 15 km/hr in still water. Stream speed is 3 km/hr. What time does it take to travel 48 km d/s?

    (a) 2 hr 20 min   (c) 3 hr 10 min   (e) 3 hr 40 min
    (b) 2 hr 40 min   (d) 3 hr 20 min

3. A person can row 3/4 km against the stream in 12 min; and along the stream in 9 min. What is the speed of the boat in still water?

    (a) 4 km/hr   (c) $4\frac{5}{8}$ km/hr   (e) 5 km/hr
    (b) $4\frac{3}{8}$ km/hr   (d) $4\frac{3}{4}$ km/hr

4. If a boat goes 4 km in 15 min along the stream, and if the speed of the stream is 2 km/hr. How much time does it take to go 4km u/s?

    (a) 10 min   (b) 20 min   (c) 30 min   (d) 45 min   (e) 75 min

5. The u/s speed of a boat is 9 km/hr. d/s speed = 14 km/hr. what is the speed of the stream?

    (a) 2 km/hr.   (c) 2.5 km/hr.   (e) 3.1 km/hr.
    (b) 2.3 km/hr.   (d) 2.8 km/hr.

6. Speed of a boat in still water is 8 km/hr. Speed of the boat u/s is 5.5 km/hr. What is its speed along the current?

   (a) 8 km/hr. (c) 9.65 km/hr. (e) 10.75 km/hr.
   (b) 8.95 km/hr. (d) 10.50 km/hr.

7. A boat takes 5hr to go 40 km along the stream. It takes 5 hr to go 20 km opposite the stream. What is the speed in still water?

   (a) 4 km/hr. (c) 8.4 km/hr. (e) 12.5 km/hr.
   (b) 6 km/hr. (d) 10.2 km/hr.

8. A motor boat whose speed is 12 km in still water goes 20 km d/s and back to the starting point in $3^5/_9$ hours. What is the speed of the stream?

   (a) 1 km/hr (c) 2 km/hr (e) 3 km/hr
   (b) 1.5 km/hr (d) 2.5 km/hr

9. A person can row a boat at 6 km/hr. in still water. If the stream level is 2 km/hr. and if it takes 1.5 hrs to row from P to Q and back. What is the distance PQ?

   (a) 1 km (b) 1.5 km (c) 2.75 km (d) 3.25 km (e) 4 km

10. The still water speed of a boat is 12 km/hr. The boat can travel 24 km against the stream and 48 km along stream in the same duration. What is the speed of the stream?

    (a) 3 km/hr. (c) 5.2 km/hr. (e) 7.75 km/hr.
    (b) 4 km/hr. (d) 6.5 km/hr.

11. The speed of a boat in still water is 10 km/hr. It takes 1.5 times as much time to row u/s compared to row the same distance d/s in the river. The speed of the current is

    (a) 2 km/hr. (c) 3.25 km/hr. (e) 5.8 km/hr.
    (b) 2.75 km/hr. (d) 4.5 km/hr.

12. A boats man can row 3 km with the stream in the same time as 2 km opposite the stream. Further, he takes 5 hours to go 24 km u/s and back, what is the speed of the current?

    (a) 2 km/hr.   (c) 3.25 km/hr.   (e) 5.8 km/hr.
    (b) 2.75 km/hr. (d) 4.5 km/hr.

13. The speed of a boat in still water is 10 km/hr, speed of current is 2 km/hr. What time does it take to go 24 km u/s and back?

    (a) 2.8 hours   (d) 5 hours
    (b) 3.9 hours   (e) 6 hours
    (c) 4.2 hours

14. The speed of a boat in still water is 10 km/hr. Speed of current is 2 km/hr. What is the distance it can travel u/s in 15 min?

    (a) 2km   (c) 2.5km   (e) 4.7km
    (b) 2.2km (d) 3.3km

Solutions

1. Let 'u' denote the speed of the boat in still water, and 'v' be the speed of the river water, then the speed of the boat d/s = u + v = (8 + 2) and its speed u/s = u - v = 8 - 2 = 6 km/hr.; If "d" is the distance x y, in km then
$$\left\{ \frac{dkm}{10\frac{km}{hr}} + \frac{d}{6} hr \right\} \text{ or } d\frac{16}{60} = \frac{4d}{15} = 4hrs$$

    $\therefore$ d = 15 km

2. s = 15 $\frac{km}{hr}$; c = $\frac{3km}{hrs}$

    $\therefore$ d = s + c = 18 km/hr.

    Time = $\frac{48km}{18\frac{km}{hr}} = \frac{8}{3}$ = 2 hr. 40 min

3. u = $\frac{\frac{3}{4}km}{\frac{12}{60}hr} = \frac{3}{4} \times \frac{60}{12} = \frac{15}{4}$ km/hr;

150   Quantitative Aptitude

$$d = \frac{\frac{3}{4}km}{\frac{9}{60}hr} = \frac{3}{4} \times \frac{60}{9} = 5\frac{km}{hr};$$

$$u = s - c = \frac{15}{4}\frac{km}{hr}; \; d = s + c = 5\frac{km}{hr}$$

$$\therefore \; 2s = \frac{15}{4} + 5 = \frac{35}{4}\frac{km}{hr}$$

$$\therefore \; s = \frac{35}{8}\frac{km}{hr}$$

4. $d = s + c = \dfrac{4km}{\frac{15}{60}hrs} = \dfrac{4 \times 60}{15} = 16\dfrac{km}{hrs}; \; c = \dfrac{2km}{hr}$

$$\therefore \; s = \frac{14km}{hrs}$$

$\therefore \; u = s - c = 12$ km/hr.

$$\therefore \; \text{Time} = \frac{4km}{12\frac{km}{hr}} = \frac{1}{3}hr = 20\min$$

5. $u = s - c = 9$ km/hr. $\therefore \; 2s = 23$ km/hr.; $c = 2.5$ km/hr.

   $d = s + c = 14$ km/hr. $\therefore \; 2c = 5$ km/hr.

6. $s = 8$ km/hr.; $u = s - c = 5.5$ km/hr. $c = 2.5$ km/hr.

   $\therefore \; s + c = 10.5$ km/hr.

7. $d = s + c = \left(40\dfrac{km}{5hr}\right) = 8\dfrac{km}{hr}; \; u = s - c = \dfrac{20km}{5hr} = \dfrac{4km}{hr}$

$$\therefore \; 2s = 12\frac{km}{hr}$$

$$\therefore \; s = \frac{6km}{hr}$$

8. Let t_1 hr and t_2 hr be the times in which the boat goes 20km d/s and u/s respectively

   $u = s - c = 12$ c, $d = s + c = 12 + c$; and $t_1 + t_2 = \dfrac{32}{9}$ hrs; and

$d = s + c = \dfrac{20 km}{t_1\, hr} = 12 + c$ (i)

and $12 - c = \dfrac{20 km}{t_2 hr} = \dfrac{20 km}{\left(\dfrac{32}{9} - t_1\right) hr}$ (ii)

(ii) $12 t_1 + c t_1 = 20$ km from (i)

$12 \times \dfrac{32}{9} - \dfrac{32}{9} c - 12 t_1 + c t_1$ (ii) $\Rightarrow 24 t_1 + \dfrac{32}{9} c = \dfrac{128}{3}$ (v);

also

$t_1 = \dfrac{20}{12 + c}$ from (i)

Then from (v) $24\left(\dfrac{20}{12+c}\right) + \dfrac{32}{9} c = \dfrac{128}{3} \Rightarrow \dfrac{480}{12+c} + \dfrac{32}{9} c$

$= \dfrac{128}{3} \Rightarrow \dfrac{15}{12+c} + \dfrac{c}{9} = \dfrac{4}{3} \Rightarrow \dfrac{135 + 12c + c^2}{9(12+c)} = \dfrac{4}{3}$

$\Rightarrow 405 + 36c + 3c^2 = 432 + 36c \Rightarrow c^2 = \dfrac{27}{3} = 9$

$\therefore c = \dfrac{3 km}{hrs}$

9. $s = 6$ kmph; $c = 2$ kmph; $u = 6 - 2 = 4 \dfrac{km}{hr}$; $d = 6 + 2 = \dfrac{8 km}{hrs}$; If x km is the distance PQ;

$PQ = \dfrac{x}{4} \dfrac{km}{\dfrac{km}{hr}} + \dfrac{x\, km}{8 \dfrac{km}{hr}} = \dfrac{3x}{8} hr = \dfrac{3}{2} hr$ (given data)

$\therefore x = 4$ km

10. $s = 12 \dfrac{km}{hr}$; Let c be the speed of the current; $u = s - c = 12 - c$; $d = 12 + c$; in say t hrs. It can travel 24 km with a speed of u.

$\therefore 24 = (12 - c) t$ and 48 km $= (12 + c) t$.

Hence $2 = \dfrac{12 + c}{12 - c} \Rightarrow 24 - 2c = 12 + c$

$\therefore 3c = 12$

$c = 4$ km/hr.

152 Quantitative Aptitude

11. $s = 10 \frac{km}{hr}$; $c = ?$ $\left. \begin{array}{l} u = s - c, \\ = 10 - c \end{array} \right\}$ $\left. \begin{array}{l} d = s + c \\ 10 + c \end{array} \right\}$ Let x be the

distance; given $\dfrac{x}{(10-c)} = \dfrac{3}{2} \dfrac{x}{10+c}$ $\Rightarrow 20 + 2c = 30 - 3c$

$\Rightarrow 5c = 10$

$\therefore$ c = 2 km/hr.

Speed c/s = u = s - c ; speed d/s d = s + c

12. $\dfrac{3}{s+c} = \dfrac{2}{s-c}$ $\Rightarrow 3s - 3c = 2s + 2c \Rightarrow s = 5c \rightarrow$ (i); further,

$\dfrac{24}{s-c} + \dfrac{24}{s+c} = 5 \Rightarrow \dfrac{24}{4c} + \dfrac{24}{6c} = 5$

$\Rightarrow \dfrac{6}{c} + \dfrac{4}{c} = \dfrac{10}{c} = 5$

$\therefore$ c = 2 km/hr.

13. s = 10 km/hr.   c = 2 km/hr.

$s - c = \dfrac{8km}{hrs}$; $s + c = 10 + 2 = \dfrac{12km}{hrs}$

$\therefore$ Time required $= \dfrac{24}{8} + \dfrac{24}{12} = 5hr$

14. $s = 10 \dfrac{km}{hr}$; $c = 2\dfrac{km}{hr}$; u = s - c; d = s + c

$s - c = 10 - 2 = 8\dfrac{km}{hrs}$

It can travel $\dfrac{8km}{60} \times 15 = 2k$ m

In 15 min

Required s; c = 2 km/hr.

Key

1)	d	2)	b	3)	b	4)	b	5)	c	6)	d	7)	b
8)	e	9)	e	10)	b	11)	a	12)	a	13)	d	14)	a

21

Geometry - Plane Figures

We consider two dimensional plane, areas.

- **Triangles:** Sum of the three interior angles = 180° (π radians) (where $\pi \simeq \frac{22}{7}$)

- Classification: Scalene Isosceles, Equilateral

- Sum of any two sides > third side

- Difference of length of any two sides < third side

- Sum of the exterior
  Angles = $\angle ACx + \angle BAy + \angle CBz = 360° = 2\pi$ rad

- Longest side will be opposite to largest angle; shortest side will be opposite to smallest angle.

- Pythagorean triangle:
  Ratio of sides =
  i) 1, 1, $\sqrt{2}$
  ii) 1, 2, $\sqrt{5}$
  iii) 3, 4, 5
  iv) 5, 12, 13
  v) 8, 15, 17....
  vi) 7, 24, 25 examples - In a 45°, 45°, 90° isosceles right triangle, the sides will be s, s, $\sqrt{2}$ s (ratio = 1:1 $\sqrt{2}$)

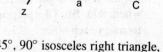

In a 30°, 60°, 90° right triangle the sides will be 's' opposite smallest $\angle 30°$, $s\sqrt{3}$ & '2s' (hypotenuse)

- Similar triangles
    - i) Corresponding angles are equal
    - ii) Corresponding lengths of sides have the same ratio

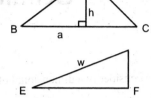

- Area of a triangle - is the number of square units enclosed.
- Area = $\left[\dfrac{1}{2}(\text{base BC} \times \text{altitude h})\right]$
- Perimeter : $(BC + CA + AB) = (a + b + c) = 2s$
- Area of a triangle = $\sqrt{s(s-a)(s-b)(s-c)}$
- Area of an equilateral triangle: here $a = b = c = a$ also each angle = 60° if "h" is AD

$\sin 60 = \dfrac{h}{a}$; also from $\triangle ABD$

$\therefore a^2 = h^2 + \left(\dfrac{a}{2}\right)^2 \Rightarrow \dfrac{3a^2}{4} = h^2$

$\therefore \triangle$, area of $\triangle ABC$ (when equilateral)

$= \dfrac{1}{2}$ base × altitude

$= \dfrac{1}{2}ha = \dfrac{\sqrt{3}}{2.2}a^2 = \dfrac{\sqrt{3}}{4}a^2$

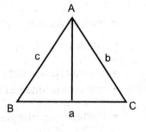

- (IMP. NOTE: Trigonometry need not be known in case of SAT, GRE, GMAT) examinations)

- **Incircle** of an **equilateral** triangle

    I is the centre of the incircle. It is the point where AD, BE, CF (perpendicular to sides) intersect each other.

    ID will be $r = \dfrac{1}{3}AD = \dfrac{1}{3}h = \dfrac{1}{3}\left(\dfrac{\sqrt{3}a}{2}\right) = \left(\dfrac{a}{2\sqrt{3}}\right)$

    (Equilateral $\triangle$'le)

I is the centroid (where the medians meet). The centroid divides each of the medians in the ratio 2:1

This is the radius of the incircle of an **equilateral** triangle.

- **Circum circle** is the circle passing through A, B, C same centre, its radius, circum radius

$$= AI = \frac{2}{3}AD = \frac{2}{3}h = \frac{2}{3}\frac{\sqrt{3}a}{2} = \frac{a}{\sqrt{3}}$$

- r the radius of the incircle $r = \frac{a}{2\sqrt{3}}$ &

Since Δ the area of an equilateral triangle $= \frac{\sqrt{3}}{4}a^2$;

$a^2 = \left(\frac{4\Delta}{\sqrt{3}}\right)$ & since $r = \frac{a}{2\sqrt{3}}$;

$\therefore a^2 = 4.3r^2 = 12r^2 = \frac{4\Delta}{\sqrt{3}}$

$\therefore \Delta = \frac{\sqrt{3}a^2}{4}$; also in the case of equilateral triangle

$2s = a + b + c = 3a$

$\therefore$'s' $= \frac{3a}{2}$

$\therefore rs = \frac{a}{2\sqrt{3}} \cdot \frac{3a}{2} = \sqrt{3}\frac{a^2}{4} = \Delta$

Thus we have the relation between Δ, r & s

- $r = \frac{\Delta}{s}$ for equilateral triangle

- The centroid I divides each median in the ratio 2:1 (as already mentioned)

- **Isosceles triangle:** When two sides have equal length s

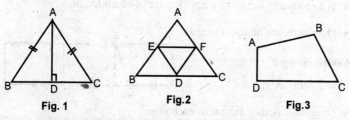

Fig. 1   Fig.2   Fig.3

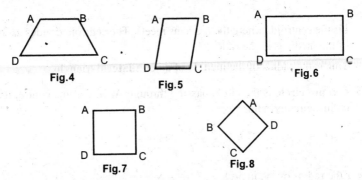

- In an isosceles Δ the altitude from the vertex A bisects the base BC. (fig 1)

- The median of a Δle divides it into two Δ les with equal area.

- The area of the triangle formed by joining the mid points of the sides of any given Δ'le is $\frac{1}{4}$ (area of the Δle) fig (ii)

- **QUADRILATERALS:** fig 3.

- **T Trapezoid:** if two sides are parallel (11^{al}) fig (4)

- **PARALLELOGRAM:** Opposite sides are 11^{al}; are equal (fig 5)

- Each Diagonal divides the area into two equal parts

RECTANGLE:

- Is a parallelogram with all angles being 90° each
- Diagonals bisect each other, (fig 6)
- Diagonals are of equal length.

SQUARE: fig 7

- Is a rectangle, adjacent sides also are of equal length
- Diagonals are of equal length.

RHOMBUS: fig 8

- is a parallelogram
- Diagonals perpendicular to each other

- Diagonals not equal in length

- Parallelograms & rectangles between the same parallel lines and with equal bases ($D_6C_6 = D_7C_7$) have equal areas.

- All the parallelograms with same base & height will have equal areas, the same as the rectangle with same base & height.

  Since area of rectangle = $D_7C_7 \times h$ & area of parallelogram

  $2 \text{ (area } B_6 D_6 C_6) = 2 \times \left(\dfrac{1}{2} \times D_6C_6 \times B_6.C_6 \sin\theta \right)$

  $= D_6C_6 \times \bar{h}$ = area of the rectangle

- **Trapezoid:** Area = $\dfrac{1}{2}(b_1 + b_2)h$

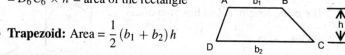

- Area of the rhombus if the diagonals are d_1 & d_2;

  Area = $\dfrac{1}{2} d_1 d_2$

  $AC = d_1$ $BD = d_2$

  $= 2 \dfrac{d_1 d_2}{4} = \dfrac{d_1 d_2}{2}$

  [∴ area = $\left(2 \times \dfrac{1}{2} \dfrac{d_2}{2} . \dfrac{d_1}{2} + 2 \times \dfrac{1}{2} \times \dfrac{d_1}{2} . \dfrac{d_2}{2} \right)$]

  Perimeter of a rectangle with length l & breadth b = 2(l+b)

  Its area = b l

- Area of a square with side a = a^2 if its diagonal = d

  Since $d^2 = 2a^2$

  ∴ area = $\dfrac{d^2}{2}$

Circle

- If radius is r, area = πr^2 : $\left(\pi = \dfrac{22}{7} \right)$

- Perimeter P = $2\pi r$

- Arc length s subtending an angle θ radians at the centre $= r\theta$ also $2\pi r = P$

$$\therefore S = \frac{P}{2\pi}\theta = \frac{P\theta}{360°}$$

- Area of the $\triangle OAB <$ area of the sector $OAB = (\frac{1}{2}r$ arc length$)$

$$= \frac{1}{2}rs = \frac{1}{2}r\frac{P\theta}{360} = \frac{1}{2}r.\frac{2\pi r\theta}{360} = \pi\frac{r^2\theta}{360}$$

- Area of circumscribing square $4r^2$

If s is arc length, r is radius area of a sector

$$= \frac{1}{2}r \times s = \frac{1}{2}r^2\theta$$

Exercises

1. If 2 & 7 are two sides of a triangle, & the third side is 'a', how many integral values are possible for 'a'?

    (a) 0   (b) 1   (c) 2   (d) 3

2. If the lengths of two sides of an isosceles triangle are 4 & 10, what is the length of the third side?

    (a) 4   (b) 6   (c) 10   (d) 14

3. 5, 7, x are the integral lengths of the sides of $\triangle PQR$. What is the least possible perimeter of the triangles?

    (a) 15 cm   (b) 18 cm   (c) 21 cm   (d) 23 cm

4. The integral values of the sides of a triangle have a product of 105. What is its perimeter?

    (a) 15 cm   (b) 17 cm   (c) 19 cm   (d) 21 cm

5. The cost of cultivating a field is Rs. 360/- @ the rate of 40 $\frac{Rs.}{ha}$. The field is triangular in shape with its base being (2 × its height). What is its base?

   (a) 500 m   (b) 600 m   (c) 700 m   (d) 900 m

6. In the △ ABC $\lfloor A = \lfloor B = 60°; AB = \sqrt{13}$ cm. The lines AD & BD intersect at D; $\lfloor D = 90°$. What is AD if DB = 2 cm?

   (a) 3 cm   (b) 3.5 cm   (c) 4 cm   (d) 4.7 cm

7. In △ ABC, AB = AC; AD, AE are drawn at 15° to AB, AC respectively to cut BC at D, E. CF is drawn at 45° to AC. AF is drawn parallel to BC. What is $\angle AFC$?

   (a) 15°   (b) 30°   (c) 55°   (d) 75°

8. The ratio of the bases of △ ABC & DEF is 5 : 4. The ratio of their heights is 2 : 3. What is the ratio of their areas?

   (a) 2 : 3   (b) 4 : 5   (c) 5 : 6   (d) 6 : 5

9. The base & height of a triangle are 8cm & 4cm. What is the height of a second triangle whose base is 12cm if its area is 3 times that of the first?

   (a) 6 cm   (b) 8 cm   (c) 10 cm   (d) 12 cm

10. The sides of a triangle are in the ratio 2 : 3 : $\sqrt{13}$ the area is 48 cm². What is the area of the triangle obtained by joining their mid points?

    (a) 6 cm²   (b) 10 cm²   (c) 12 cm²   (d) 24 cm²

11. The sides of a triangle are 12 cm, 22 cm & 30 cm. What is the radius of its incircle?

    (a) $\frac{5\sqrt{2}}{4}$ cm   (b) $\frac{5\sqrt{2}}{2}$ cm   (c) $\frac{5}{2}$ cm   (d) 5 cm

12. The two parallel sides are in the ratio of 3:4 in a trapezium. The perpendicular distance between them is 10 cm. What is the length of the shorter (parallel) side. If the area is 175 cm²?

    (a) 10 cm   (b) 15 cm   (c) 20 cm   (d) 25 cm

13. Each side of a rhombus is 52 cm. One diagonal is of length 96 cm. What is its area?

    (a) 240 cm²   (c) 960 cm²
    (b) 480 cm²   (d) 1920 cm²

    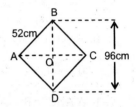

14. In a parallelogram ABCD, the perpendicular distance between AB & CD is $4\sqrt{3}$ cm. $\angle BCD = 60°$; If E is the foot of the perpendicular BE on CD; It is also given that ED = 4 cm.

    What are the lengths of its diagonals?

    (a$_1$) $4, 4\sqrt{3}$ cm   (c$_1$) $8, 8\sqrt{3}$ cm
    (b$_1$) $6, 6\sqrt{3}$ cm   (d$_1$) $12, 12\sqrt{3}$ cm

15. In a 'fashion technology' college the 'welcome' fabric banner has shrunk by 10% in length & 5% in width on blenching. What is the new area of the banner as a percentage of originally proposed area?

    (a) 69.8%   (b) 79.7%   (c) 82.5%   (d) 85.5%

16. The length of a rectangle is decreased by (r - 5) % & the breadth is increased by (r + 5) %. Find r if the area of the rectangle is unaltered?

    (a) 30.5%   (b) 32.01%   (c) 34.35%   (d) 35.45%

17. If the area of a rectangle increased by 20% while the breadth remains the same, what is the ratio of the lengths of new & old rectangles?

    (a) 3:4   (b) 4:5   (c) 5:6   (d) 6:5

18. The diagonals of a rhombus are in the ratio 3:1; its area is 150 sq cm. What is the perimeter of rhombus?

   (a) $10\sqrt{10}$ cm   (b) $15\sqrt{10}$ cm   (c) $18\sqrt{10}$ cm   (d) $20\sqrt{10}$ cm

19. The parallel sides of a trapezoidal area are in the ratio 8:5. The perpendicular distance between them is 12 m. What is the longer side if the area is 1560 m^2?

   (a) 20 m   (b) 100 m   (c) 160 m   (d) 260 m

20. A circular sector subtending 240° at the center has an area equal to $6\frac{2}{7}$ cm^2. What is the area of the circle?

   (a) $\frac{\pi}{2}$ cm^2 (b) π (c) 2π (d) 3π cm^2

Solutions

1. i) a must be $< 2 + 7$ ie 9 'a' must be $> 7 - 2$ ie 5

   Thus the possible integral values are only 6, 7, 8. Three

2. The only two possibilities are that the third side should be either 4 or 10. Let us try them.

   Is $4 < 10 + 10$?   yes so 10 is a possible solution

   Is $4 > 10 - 10$?   Yes

   Is $10 < 4 + 4$ ie 8? no $\to$ So 4 cannot be a solution

   Is $10 > 4 - 4$ ie 0? yes

   So the only possibility is 10 the other possible alternative answer choices (6, 14) can be eliminated straight away, since they are given as trap answers (one is the difference & the other the sum of the two given values 4 & 10). Infact since the said triangle is described an isosceles one it cannot have any other values than 4 or 10.

3. The lengths should satisfy the following inequalities

   i) $5 < 7 + x \Rightarrow 7 + x > 5 \Rightarrow x > -2$ (i)

   ii) $7 < x + 5 \Rightarrow x + 5 > 7 \Rightarrow x > 2$ (iii)

   iii) $x < 5 + 7 \Rightarrow 12 > x \Rightarrow x < 12$ (iv)

   iv) $5 > 7 \sim x$

v) $7 > x \sim 5$

vi) $x > 7 - 5 \Rightarrow x > 2$ (ii)

We conclude $2 < x < 12$ (vii) is the range of possible values of x

i) is obviously satisfied. The least integral value is 3 & largest is 11 satisfying the inequality range.

$7 > 5 - 3$ or $11 - 5$

ie 2 or 6

There are satisfied by (vii)

$5 > 7 \sim 3$ or $7 \sim 11$

ie $7 - 3$ or $11 - 7$ ie 4

(This) There are satisfied

Thus least value 3 is to be accepted, and then the least perimeter will be $5 + 7 + 3 = 15$ cm.

4. The sides are 3, 5, 7 now do they satisfy the length rules of a triangle

   i) Is $3 < 5 + 7$ also iv) Is $3 > 7 - 5$
   ii) $5 < 7 + 3$?   v) Is $5 > 7 - 3$
   iii) Is $7 < 3 + 5$ vi) Is $7 > 5 - 3$

Thus (3, 5, 7) is satisfactory triad of values; perimeter

$P = 3 + 5 + 7 = 15$ cm

```
        105
        /\
       3  35
          /\
         5  7
```

5. Cost of cultivation = Rs. 360/- Rate is $\dfrac{40 \text{ Rs.}}{\text{ha}}$;

∴ Area of the triangular field = $\dfrac{360}{40} = 9 ha = 9 \times 10^4 m^2$

If b is the base, h is the height given as $\left(\dfrac{b}{2}\right)$

Area = $\dfrac{1}{2}$ base × altitude = $\dfrac{1}{2} \times b \times \dfrac{b}{2} = 9 \times 10^4 m^2$

∴ $b^2 = 36 \times 10^4 m^2$

$\Rightarrow b = 6 \times 10^2$ m = 600 m

Geometry - Plane Figures   163

6. ∠C also will be 60°. ABC will be equilateral &
   AB = $\sqrt{13}$ cm. In △ ADB; DB = 2 cm
   AB = $\sqrt{13}$ cm
   ∴ AD = $\sqrt{AB^2 - DB^2}$ = $\sqrt{13-4}$
   = $\sqrt{9}$ = 3 cm

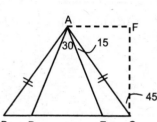

7. ∠BAC = 15° + 30° + 15° = 60°
   ∴ Since AB = AC
   ∴ ∠ABC = ∠ACB = θ say
   Now 2θ + 60° = 180°
   ∴ θ = 60°. ABC is equilateral

   Since ∠ADC = ∠ABC + ∠BAD = 60° + 15° = 75°
   ∠ACB = 60°; ∠ACF = 45°
   =105° (as it should be)
   DC ∥ AE; ∠AFC = 180° − ∠DCF
   = 180° - (60°+45°) = 75°

8. $a_1 = \frac{1}{2} b_1 h_1$; $a_2 = \frac{1}{2} b_2 h_2$

   ∴ $\frac{a_1}{a_2} = \frac{1/2 \, b_1 h_1}{1/2 \, b_2 h_2}$

   $\frac{5}{4} \cdot \frac{2}{3} = \frac{5}{6}$

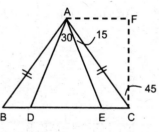

9. $\frac{a_2}{a_1} = \frac{1/2 \, b_2 h_2}{1/2 \, b_1 h_1} \Rightarrow 3 = \frac{12 h_2}{8 \times 4}$

   ∴ $h_2 = \frac{3 \times 8 \times 4}{12}$ = 8 cm

10. Let BC: BA: CA be 3x: 2x: $\sqrt{13}x$
    ABC is a right triangle

∵ $9x^2 + 4x^2 = 13x^2$

If D, E, F are the midpoints then DE||AB; EF||BC; & FD||AC; Δ EDF is a right triangle similar to Δ ABC

$DE = \frac{1}{2}AB = x; EF = \frac{1}{2}BC = \frac{3}{2}x; DF = \frac{1}{2}AC = \frac{\sqrt{13}}{2}x$

Area of Δ DEF = $\frac{1}{2}$ DE.EF

$= \frac{1}{2}x \cdot \frac{3}{2}x$

$= \frac{3}{4}x^2$

Now, Given $\frac{1}{2}$ AB. BC

$= \frac{1}{2} \cdot 3x \cdot 2x$

$= 3x^2 = 48\ cm^2$

∴ $x^2 = 16$

∴ $x = 4$

∴ Area of Δ DEF = $\frac{3}{4} \cdot 4^2$

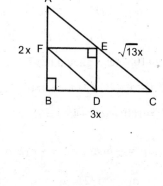

$= 12\ cm^2$

$= \left(\frac{1}{4} \Delta ABC\right)$

11. Let a, b, c be its sides, 2s its perimeter; r its radius of in circle & Δ its area 2s = a + b + c = 12 + 22 + 30 = 64 cm

∴ s = 32 cm

∴ (s - a) = 20 cm; (s - b) = 10 cm; (s - c) = 2 cm;

∴ Area Δ = $\sqrt{s(s-a)(s-b)(s-c)}$

$= \sqrt{32 \times 20 \times 10 \times 2}$

$= \sqrt{64 \times 200}$

Geometry - Plane Figures 165

$$= 80\sqrt{2}\, cm^2$$

Radius of the incircle $r = \dfrac{\Delta}{s} = \dfrac{80\sqrt{2}}{32} = \dfrac{5}{2}\sqrt{2}\ cm$

12. Let 3x, 4x be the lengths of the parallel sides AB, CD; & h = 10 cm;

Area of the trapezium = area of △ ABD area of △ BDC

$$= \dfrac{1}{2}(3x)h + \dfrac{1}{2}(4x)h$$

$$= \dfrac{7xh}{2} = 175\ cm^2$$

$$\therefore \dfrac{7x}{2} \cdot 10 = 175$$

$$\therefore x = \dfrac{350}{70} \Rightarrow 5$$

∴ Shorter required side = 15 cm

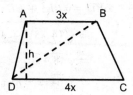

13. $AO = \sqrt{AB^2 - BO^2}$

$$= \sqrt{52^2 - 48^2}$$

$$= \sqrt{(4 \times 13)^2 - (4 \times 12)^2} = \sqrt{(4 \times 5)^2} = 20\ cm$$

∴ AC = 40 cm; BD = 96 cm

∴ Area of the rhombus = $\dfrac{1}{2} \times 96 \times 40 = 1920$ sqcm

14. Draw BE, DF to DC & AB.

△ ECB is a 30° - 60° - 90° right triangle (since ∠BCD = 60°) & BE = $4\sqrt{3}$ cm (i)

∴ CE = 4 cm   ii) & (iii) hypotenuse BC = 2 × 4 = 8 cm (iii)

(check $CE^2 + BE^2 = 4^2 + \left(4\sqrt{3}\right)^2 = 16 + 48 = 64$

By Pythagoras theorem this will be BC^2

∴ BC = $\sqrt{64}$ = 8 cm

∴ AD will also be 8 cm. Also ED = 4 cm (given)

Then CD = CE + ED = 4 + 4 = 8 cm (iv)

Also AB will be equal to CD = 8 cm (v)

Then ABCD will be a special parallelogram where all the sides are equal. i.e it is a rhombus side 8 cm (perimeter = 32 cm)

Join BD in $\triangle$ BDE

∴ BE = $4\sqrt{3}$ cm & DE = 4 cm

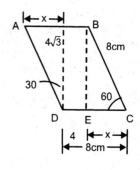

∴ Diagonal BD of the right triangle

BED = $\sqrt{BE^2 + ED^2}$

$= \sqrt{(4\sqrt{3})^2 + 4^2}$

$= \sqrt{48 + 16}$

$= \sqrt{64} = = 8$ cm

(ie 8 cm = BD) (vi)

Now consider the hypotenuse AE of the right triangle ABE;

$AE^2 = AB^2 + BE^2$

$= (x + 4)^2 + (4\sqrt{3})^2$

$= 8^2 + (4\sqrt{3})^2$

[(note AB = BF + FA= 4 + FA) Also $\angle BAD = \angle BCD = 60°$;]

= 64 + 48 = 112

∴ AE = $\sqrt{112} = 4\sqrt{7}$ cm

(i.e AE = $4\sqrt{7}$ cm (vii))

Also $\angle BAD = \angle BCD = 60°$;

∴ $\angle ADF$ (in $\triangle$ ADF) = 30°

∴ ADF will also be a 30 - 60 - 90° right triangle since AD = BC = 8 cm (vii)

∴ FA the side opposite 30° will be $\frac{1}{2} \times$ hypotenuse

$= \frac{1}{2} \times 8 = 4$ cm

[∴ BA = 4 + 4 = 8 cm]

Now the diagonals of the rhombus ABCD are AC & BD;

BD = 8 cm   (vi)

Area of the rhombus (parallelogram) = (DC × BE) = base × alt)

= 8 × 4$\sqrt{3}$ from (iv) and (i)

From (iii) and

= 32 $\sqrt{3}$ cm²

Then $\left[\frac{1}{2} \text{ product of diagonals}\right]$

$\frac{1}{2}$ AC. BD i.e $\left(\frac{1}{2} \text{AC. 8}\right)$ should be $32\sqrt{3}$

∴ $AC = 32\sqrt{3} \times \frac{2}{8} = 8\sqrt{3}$ cm

The diagonals are → BD, AC are → 8cm & $8\sqrt{3}$ cm   (c_3)

(viii) (from (v) & (vii)) perimeter = 4 × 8 = 32 cm

[Answer choices

(a_2) 16 cm   (b_2) 24 cm   (c_2) 32 cm   (d_2) 40 cm

(a_3) 4, $4\sqrt{3}$ cm   (b_3) 6, $6\sqrt{3}$ cm   (c_3) 8, $8\sqrt{3}$ cm   (d_3) 12, $12\sqrt{3}$ cm]

15. $\frac{90}{100} l \times \frac{95}{100} b = \frac{8550}{10,000} lb = 0.855$ lb

∴ area of the banner = $\frac{855}{1000} \times 100$

= 85.5% of the proposed area of the farmer

16. $\left[l - \frac{r-5}{100}l\right] \left[b + \frac{r+5}{100}b\right] = lb$ (given)

⇒ (100 - r + 5)(100 + r + 5) = 10,000

⇒ (105 - r)(105 + r) = 10,000 .

$\Rightarrow \quad 105^2 - r^2 = 100^2$

$\Rightarrow \quad r^2 = 105^2 - 100^2$

$\qquad = (105+100)(105-100)$

$\qquad = 205 \times 5$

$\qquad = 1025$

$\therefore \quad r^2 = 1025$

$\therefore \quad r = 32.01$

Check $\left[l - \dfrac{27.01l}{100}\right]\left[b + \dfrac{37.01b}{100}\right]$

$\Rightarrow lb\left(\dfrac{7.299}{100} \times \dfrac{137.01}{100}\right) = lb = \dfrac{10,000.3599}{10,000} \simeq 1$

17. $\dfrac{(1.2l)b}{lb} = 1.2;$

Or 12:10 Or 6:5

18. Let d_1, d_2 be the diagonals AC, BD,

    $AC = d_1 = 3d_2$ (given)

    Area $= \dfrac{1}{2}d_1 d_2$

    $\qquad = \dfrac{1}{2} \times 3d_2 \times d_2$

    $\qquad = 150$ cm^2

    $\therefore d_2^2 = \dfrac{300}{3} cm^2$

    $\therefore d_2 = \sqrt{100} = 10$ cm

    $\therefore d_1 = 30\ cm$

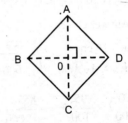

In Δ AOD;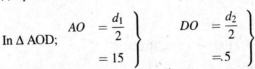

∴ AD = $\sqrt{15^2 + 5^2}$

= $\sqrt{225 + 25}$

= $\sqrt{250}$

= $5\sqrt{10}$ cm

∴ Perimeter of rhombus = 4AD = $20\sqrt{10}$ cm

19. Let 8x, 5x be the sides DC & AB

Area = $\frac{1}{2}(8x + 5x) \times 12 = 1560\ m^2$

∴ $13x = \frac{2 \times 1560}{12} = 260\ m$

∴ x = 20 m

∴ DC = 8x = 160 m

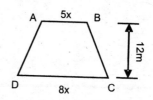

20. Area of the sector = $\frac{\pi r^2 \theta}{360}$

= $\frac{\pi r^2 240}{360}$

= $\frac{2}{3}\pi r^2$

= $\frac{44}{7} \frac{r^2}{3}$

= $\frac{44}{7}\ cm^2$ (given)

∴ $\frac{r^2}{3} = 1$ or $r = \sqrt{3}$ cm

∴ area of circle = $\pi r^2 = 3\pi\ cm^2$

Key

1)	d	2)	c	3)	a	4)	a	5)	b	6)	a
7)	d	8)	c	9)	b	10)	c	11)	b	12)	b
13)	d	14)	c	15)	d	16)	b	17)	d	18)	d
19)	c	20)	d								

22

Solid Figures

Exercises

Rectangular parallelepiped problems

1. A wooden plank 3.6 m × 2 m is floating on a lake. It sinks by 1 cm when a person gets on to it. What is his weight?

    (a) 60 kg   (b) 66 kg   (c) 72 kg   (d) 78 kg

2. If 20 people take a bath in a water tank 20 m × 10 m in size, & if on an average the displacement of water by each person is 4 m³, what is the rise in water level of the tank?

    (a) 15 cm   (b) 25 cm   (c) 30 cm   (d) 40 cm

3. Water is let into a rectangular water tank 100 m × 44 m at a speed of $\frac{8km}{hr}$ through a pipe of 3.5 cm radius. What is the rise in the water level in the tank in one hour?

    (a) 0.5 cm   (b) 0.6 cm   (c) 0.7 cm   (d) 0.8 cm

4. The edges of a cuboid are in the ratio 2:3:4. Its surface area is 468 cm². What is its volume?

    (a) 616cm³   (b) 628cm³   (c) 648cm³   (d) 650cm³

5. What is the length of the longest rod that can be placed in a room 30' long 20' broad & 15 ft high?

    (a) 30 ft   (b) 34 ft   (c) 37 ft   (d) 39 ft

6. If the areas of the three adjacent faces of a cuboidal box are 270 cm^2, 162 cm^2 & 135 cm^2 respectively. Find the volume of the box

   (a) 2370 cm^3 (b) 2400 cm^3 (c) 2430 cm^3 (d) 2460 cm^3

7. A water tank 4 m × 3 m has water upto a depth of 1.1 m. What is the wetted area?

   (a) 24 m^2 (b) 25.9 m^2 (c) 26.8 m^2 (d) 27.4 m^2

8. How many bricks measuring 25 cm × 12.5 cm × 7.5 cm are needed to construct a wall 8 m × 5 m × 22.5 cm is 5% of the volume of the wall is occupied by mortar?

   (a) 3444   (b) 3532   (c) 3648   (d) 3812

9. A water tank of capacity 10,000 litres measures externally 4.1 m × 2.6 m × 1.1 m. Its side walls are 5 cm thick. What is the thickness of the bottom?

   (a) 8 cm   (c) 20 cm
   (b) 10 cm (or) 16 m   (d) 40 cm

10. A solid block of iron 8.4 cm × 7.7 cm × 4.8 cm is moulded into a sphere. What is the radius of the sphere?

    (a) 2.1 cm   (b) 4.2 cm   (c) 8.4 cm   (d) 16.8 cm

11. The surface area of a sphere is the same as the curved surface area of a right circular cylinder with height 6 cm & radius 3 cm. What is the radius of the sphere?

    (a) 1.5 cm   (b) 3 cm   (c) 4.5 cm   (d) 6 cm

12. A metallic cone radius 10 cm, height 20 cm is melted & made into spheres radius 5 cm each. How many spheres can be made?

    (a) 4   (b) 6   (c) 8   (d) 12

Solid Figures 173

13. A semi sphere & a cone have equal base. If their heights are also equal, what is the ratio of their curved surfaces?

    (a) 1:2　　　(b) 2:1　　　(c) $1:\sqrt{2}$　　　(d) $\sqrt{2}:1$

14. A semi spherical bowl radius 6 cm contains a liquid. It is filled in cylinders 4 cm dia, 4 cm high. How many cylinders will be filled?

    (a) 3　　　(b) 6　　　(c) 9　　　(d) 12

15. What is the ratio of the curved SA'S of a cone, a semi sphere, & a cylinder with the same base radius r, height also r?

    (a) $1:\sqrt{2}:\sqrt{2}$　　(b) $\sqrt{2}:\sqrt{2}:2$　　(c) $1:2:\sqrt{2}$　　(d) $1:2:2$

| Solutions |

1. Volume displaced = $3.6 \times 2 \times \dfrac{1}{100} m^3 = \dfrac{7.2}{100} m^3$

   Weight of water displaced = $\dfrac{7.2}{100} \times 1000 \dfrac{kg}{m^3}$

   Weight of the person = 72 kg

2. Volume of water displaced = $20 \times 4 = 80 \ m^3$

   Water spread area = $20 \times 10 = 200 \ m^2$

   $\therefore$ Rise in water level = $\dfrac{80}{200} m = \dfrac{2}{5} m = 40 \ cm$

3. Incoming water in one hour = $\dfrac{22}{7} \times \dfrac{7}{2} \times \dfrac{7}{2} cm^2 \times 8km$

   $= \dfrac{22}{7} \times \dfrac{7}{2} \times \dfrac{7}{2} \times \dfrac{1}{100} \times \dfrac{1}{100} \times 8000 \ m^3 = \dfrac{22 \times 14}{10} m^3$

   This is in one hour

   Area of water tank = $100 \times 44 m^2 = 4400 m^2$

   Rise in the water tank = $\dfrac{22 \times 14}{10 \times 4400} \dfrac{m^3}{m^2}$

   $= \dfrac{22 \times 14}{10 \times 4400} \times 100 \ cm$

   $= \dfrac{7}{10} cm$ in one hour = $0.7 cm$

4. Let the sides be 2x, 3x, 4x, (not in the order of l, b, d) then surface area = $2(lb + bd + dl) = 2(6x^2 + 12x^2 + 8x^2) = 52x^2$

This is given to be 468 cm^2

$$\therefore x^2 = \frac{468}{52} = \frac{36}{4} = 9$$

x = 3 cm; so the sides are 12, 9, 6 cm

∴ Volume = 648 cm^3

5. Length of the rod = diagonal of the cuboid sides 30′, 20′, 15′.

$$= \sqrt{l^2 + b^2 + h^2}$$
$$= \sqrt{30^2 + 20^2 + 15^2}$$
$$= \sqrt{900 + 400 + 225}$$
$$= \sqrt{1525}$$
$$= 39 \text{ ft}$$

6. Let l, b, d be the dimensions of the edges.

Then $a_1 = lb = 270 \text{ cm}^2$; $a_2 = ld = 162 \text{ cm}^2$; $a_3 = bd = 135 \text{ cm}^2$

Then $(lbd)^2 = a_1 a_2 a_3$
$= 270 \times 162 \times 135$
$= 135^2 \times 4 \times 81$

Then $(lbd) = 135 \times 2 \times 9$

$$\therefore d = \frac{135 \times 2 \times 9}{270} = 9 \text{ cm}$$

$$b = \frac{135 \times 2 \times 9}{162} = 15 \text{ cm}; \; \& \; l = \frac{135 \times 2 \times 9}{135} = 18 \text{ cm}$$

Then volume of the box = $18 \times 15 \times 9 = 270 \times 9 = 2430 \text{ cm}^3$

7. = 2(4 + 3) × 1.1 m + 4 × 3

= 2 × 7 × 1.1 + 12

= 15.4m^2 + 12

= 27.4 m^2

[i.e 2lh+2bh+lb] ⇒ 27.4 m²

8. Volume of the wall = $8 \, m \times 5 \, m \times \dfrac{225}{1000} m^3 = \dfrac{9000}{1000} m^3 = 9 \, m^3$

   Mortar occupies 5% of this volume

   Remaining volume = $\dfrac{95}{100} \times 9 = 8.55 \, m^3$

   Vol of each brick = $\dfrac{25}{100} \times \dfrac{125}{1000} \times \dfrac{75}{1000} m^3$

   $= \dfrac{1}{4} \times \dfrac{1}{8} \times \dfrac{3}{40} m^3 = \dfrac{3}{1280} m^3$

   ∴ Bricks needed = $\dfrac{8.55}{3} \times 1280 = 2.85 \times 1280 = 3648$

9. Internal dimensions of the plan

   = 4.1 m - 10 cm & 2.6 m - 10 cm = 4 m & 2.5 m

   Its capacity = $4 \times 2.5 \, m \times (1.1 - x \, m)$

   $= \dfrac{10,000 \times 1000}{10^6} = 10 \, m^3$ (given)

   Or $(1.1-x) = \dfrac{10}{4 \times 2.5} m = 1 \, m$

   ∴ x = 0.1 m or 10 cm or 1 dm

10. $\dfrac{4}{3} \times \dfrac{22}{7} r^3 = 8.4 \times 7.7 \times 4.8 \, cm^3$

    ∴ $r^3 = \dfrac{84 \times 77 \times 48 \times 3 \times 7}{1000 \times 4 \times 22}$

    $= \dfrac{21 \times 7 \times 24 \times 21}{1000} = \dfrac{21 \times 21 \times 21 \times 8}{1000} = \left(\dfrac{7 \times 3 \times 2}{10}\right)^3$

    ∴ $r = \dfrac{42}{10} = 4.2 \, cm$

11. $4\pi r^2 = 2\pi \times 3 \times 6$

    ∴ $r^2 = \dfrac{2\pi \times 18}{4\pi} = 9$

    ∴ r = 3 cm

12. $\dfrac{1}{3} \pi \times 10^2 \times 20 = \left[\dfrac{4}{3}\pi \times 5^3\right] n$

$\therefore n = \dfrac{1}{3} \dfrac{\pi \times 10^2 \times 20 \times 3}{4\pi \times 125} = \dfrac{2000}{500} = 4$

13. Height of the semi sphere = r

    Its curved surface area $= 2\pi r^2$

    Curved surface of cone $\pi r l = \pi r \sqrt{r^2+h^2} = \pi r \sqrt{(r^2+r^2)} = \sqrt{2}\pi r^2$

    $\therefore$ Ratio of the curved surfaces $= 2 : \sqrt{2} = \sqrt{2} : 1$

14. Volume of the liquid $= \dfrac{2}{3}\pi r^3$

    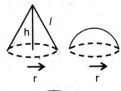

    $= \dfrac{2}{3} \times \pi \times 216$

    $= 144 \ cm^3$

    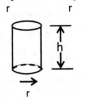

    Volume of each cylinder $= \pi \times 2^2 \times 4 = 16\pi$

    $\therefore$ h = 4 cm; r = 2 cm

    No. of cylinders $= \dfrac{144\pi}{16\pi} = 9$

15. Curved SA of cone π r l;

    $l = \sqrt{r^2+h^2} = \sqrt{r^2+r^2} = \sqrt{2}r$

    $\therefore$ SA of cone $= \pi r . \sqrt{2} r = \sqrt{2}\pi r^2$

    Curved SA of semi sphere $= 2\pi r^2$

    (height = r)

    Curved SA of cylinder $2\pi r^2 (2\pi r h = 2\pi r)$

    $\therefore$ Ratio cone, HS;

    Cylinder $= \sqrt{2}\pi r^2 : 2\pi r^2 : 2\pi r^2 = \sqrt{2} : 2 : 2 = 1 : \sqrt{2} : \sqrt{2}$

Key

1)	c	2)	d	3)	c	4)	c	5)	d	6)	c
7)	d	8)	c	9)	b	10)	b	11)	b	12)	a
13)	d	14)	c	15)	a						

23

Heights & Distances

Exercises

1. A person from the top of a hill observes a vehicle moving towards him at a uniform speed. It takes 10 minutes for the angle of depression to change from 45° to 60°. How soon after this will the vehicle reach the hill?

   (a) 12m 20s   (b) 13m   (c) 13m 40s   (d) 14m 24s

2. From a point A on ground, the angle of elevation of the top of a tower is 45°. If the tower is $100\sqrt{2}$ m in height, what is the distance from A to the foot of the tower B?

   (a) 120m   (b) 141.4m   (c) 173.2m   (d) 186.8m

3. A boat is moving away from an observation tower. It makes an angle of depression of 60° with an observer's eye when at a distance of 50 m from the tower. After 8 sec the angle of depression becomes 30°. What is the approximate speed of the boat, assuming that it is running in still water?

   (a) 33kmph   (b) 42kmph   (c) 45kmph   (d) 50kmph

4. The angle of elevation of the sun, when the length of the shadow of a tree is the same as the height of a tree is

   (a) 30°   (b) 45°   (c) 60°   (d) 90°

5. A person standing at A looks at the top of a tower T, the angle of elevation being 45°. He walks a certain distance towards the tower to

B. looks again at the top of the tower. Angle of elevation now is 60°. What is the distance from A to T?

(a) $4\sqrt{3}$ units
(b) 8 units
(c) 12 units
(d) Data in adequate

6. The angle of elevation from a ship to a light house is 45°. Another ship on the other sides makes an angle of elevation of 60°. The light house is 120 m high, what is the distance between the two ships?

(a) 40m
(b) $40\sqrt{3}$
(c) 120
(d) $(120 + 40\sqrt{3})$ m

Solutions

1. Let h be the height of the hill

$$\frac{h}{x+y} = \tan 45 = 1$$

$\therefore h = x+y$ (i)

$$\frac{h}{y} = \tan 60 = \sqrt{3}$$ (ii)

$\therefore h = x+y = y\sqrt{3}$

$\therefore x = 0.732 y$

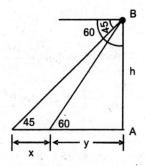

Now it takes 10 min to cover distance x m

$\therefore$ In 1 min the vehicle overs 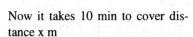 $\frac{x}{10}$m → 1 min

To cover y it then takes → 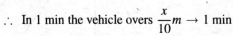 $\frac{1}{\left(\frac{x}{10}\right)} y = \frac{10y}{x}$ min

$= \frac{10y}{0.732y} = \frac{10}{0.732}$ min

$= 13.66$ min

i.e **13 min 40 sec**

180 *Quantitative Aptitude*

2. $\tan 45 = \dfrac{BC}{AB}$

   $= \dfrac{100\sqrt{2}}{AB} = 1$

   $\therefore AB = 100\sqrt{2} = 100 \times 1.414$

   $\qquad\qquad = 141.4 m$

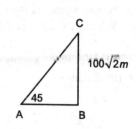

3. $\dfrac{h}{50} = \tan 60;$

   $\dfrac{h}{(x+50)} = \tan 30$

   $\therefore h = 50\sqrt{3}; h = (x+50)\dfrac{1}{\sqrt{3}}$

   $\therefore 50\sqrt{3} = \dfrac{(x+50)}{\sqrt{3}}$

   $\therefore 150 = x + 50$

   $\therefore x = 100 m$

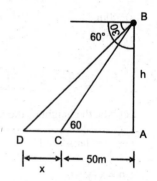

   Speed of the boat $= \dfrac{100m}{8\sec} = \dfrac{100}{8} \times \dfrac{60 \times 60}{1000} = \dfrac{360}{8} =$ **45kmph**.

4. $\dfrac{BC}{AB} = \dfrac{h}{h} = 1 = \tan\theta$

   $\therefore = 45^0$

5. $\dfrac{h}{AT} = \tan 45 = 1$

   $\therefore h = (x+y);$

   $\dfrac{h}{BT} = \tan 60 = \sqrt{3}$

   $\therefore h = BT\sqrt{3} = y\sqrt{3}$

   $\therefore x+y = y\sqrt{3}$ or $x =$ **0.732y**

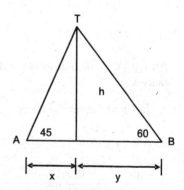

   We can not determine x + y since y is not known

6. $\dfrac{h}{x} = \tan 45 = 1$

∴ $h = x = 120$ m and $\dfrac{h}{y} = \tan 60 = \sqrt{3}$

∴ $y = \dfrac{h}{\sqrt{3}}$

∴ $(x+y) = 120 + \dfrac{120}{\sqrt{3}}$

$= (120 + 40\sqrt{3})$ m

Key

| 1) | c | 2) | b | 3) | c | 4) | b | 5) | d | 6) | d |

24

Algebra

A. Linear Equations

$p_1 x + q_1 = 0$ is an example of a linear equation in one variable. $p_2 x + q_2 y + r = 0$ is a linear equation of two variables x and y.

Two linear equations $p_2 x + q_2 y + r = 0$ and $p_3 x + q_3 y + r_3 = 0$ are said to be 'consistent' if they have at least one set of solutions. They are 'inconsistent' if they have no solution.

a) They have a unique solution if $\dfrac{p_2}{p_3} = \dfrac{q_2}{q_3}$

b) They have no solution if $\dfrac{p_2}{p_3} = \dfrac{q_2}{q_3} = \dfrac{r_2}{r_3}$

c) They have an infinite no of solutions if $\dfrac{p_2}{p_3} = \dfrac{q_2}{q_3} = \dfrac{r_2}{r_3}$

Homogeneity: The system of equations $p_2 x + q_2 y = 0$ and $p_3 x + q_3 y = 0$ have

a) the only solution $x = 0$ and $y = 0$ when $\dfrac{p_2}{p_3} = \dfrac{q_2}{q_3}$

b) they have an infinite no of solutions when $\dfrac{p_2}{p_3} = \dfrac{q_2}{q_3}$

Algebra 183

Exercises

1. Given $6x + 3y = 21$ (i) and $6x - 7y = -9$ (ii) find $10y$;

    (a) 10   (b) 20   (c) 30   (d) 40

2. 100 apples and 50 mangoes cost Rs.800/-. 50 apples and 100 mangoes cost Rs.850/- What is the difference of the costs of an apple and a mango?

    (a) 0   (b) 1   (c) 2   (d) 3

3. Solve $3x + 4y = -6$ (i) and $5x - 2y = 16$ (ii)

    (a) (-2, -3)   (b) (2, -3)   (c) (-2, 3)   (d) (2, 3)

4. 4 radios and 2Tv's cost Rs.42000/- The difference of their costs Rs.19500/-. What is the cost of a TV?

    (a) Rs.1000/-   (b) Rs.20000/-   (c) Rs.30000/-   (d) Rs.40000/-

Solutions

1. Subtracting (ii) from (i) $10y = 30$

2. Let a and n be the cost in rupees of an apple and a mango.

    Given $100a + 50m = Rs.800/-$ (i)

    $50a + 100m = Rs.850/-$ (ii)

    Multiplying (ii) by 2 = $100a + 200m = Rs.1700/-$ (iii)

    Subtracting (i) from (iii) $150m = Rs.900/-$

    $\therefore m = Rs.6/-$

    From i then $100a = 800 - 300 = 500$

    $\therefore a = Rs.5/-$

    Reqd difference $m - a = Re 1/-$ ans

3. Multiply (i) by 5: 15x + 20y = - 30 (iii)
   Multiply (ii) by 3: 15x - 6y = - 48 (iv)
   Subtracting (iv) from (iii): 26y = - 78   ∴ y = - 3
   Substituting in (i) 3x = - 6 + 12 = 6   ∴ x = 2
   Check: LHS of (ii) 5x - 2y = 10 + 6: 16 = RHS   QED

4. Let t,r be the costs
   $2t + 4r = 42000$ (i)
   $t - r = 19500$ (ii)
   from(i) $t + 2r = 21000$ (iii)
   from (iv) $2t - 2r = 39000$ (iv)
   ──────────────────────
   add $3t = 60000$
   ∴ $t = Rs.2000/-$

B. Quadratic Equations

$ax^2 + bx + c = 0$ is a quadratic equation Its roots are α, and

$$\beta = \left[\frac{-b \pm \sqrt{b^2 - 4ac}}{2a}\right]$$

1. The roots are 'imaginary' if the 'discriminant' $D = \sqrt{b^2 - 4ac}$ is negative

2. They will be 'real' and 'equal' when $D = 0$ and the roots are $\left(-\frac{b}{2a}\right)$

3. They will be 'irrational' and 'unequal' when $D > 0$ and is a perfect square

4. They will be 'irrational' and 'unequal' when $D > 0$ and D is not a perfect square

The square roots have a sum $= (\alpha + \beta) = \left(\frac{-b}{a}\right)$ and a product
$\alpha\beta = \left(\frac{c}{a}\right)$

Exercises

1. If α, β are the roots of $x^2 - bx + c = 0$ (i) find $\alpha^4 - \beta^4$

   (a) $b\sqrt{b^2 - 4c}\,(b^2 + 2c)$
   (b) $b\sqrt{b^2 + 4c}\,(b^2 + 2c)$
   (c) $b\sqrt{b^2 - 4c}\,(b^2 - 2c)$
   (d) $b\sqrt{b^2 + 4c}\,\sqrt{b^2 - 2c}$

2. Solve $9x^2 + 15x + 4 = 0$

   (a) $\dfrac{-1}{3}, \dfrac{-4}{3}$
   (b) $\dfrac{1}{3}, \dfrac{-4}{3}$
   (c) $\dfrac{-1}{3}, \dfrac{4}{3}$
   (d) $\dfrac{1}{3}, \dfrac{4}{3}$

3. If $3 + I\sqrt{5}$ is a root of the equation $x^2 + mx + n = 0$ m, n are real. Find m and n

   (a) -6, -14
   (b) 6, -14
   (c) -6, 14
   (d) 6, 14

4. α, β are the roots of the equation $3x^2 - 8x + 4 = 0$, find $\alpha^2 - \beta^2$

   (a) $\dfrac{-32}{9}$
   (b) $\dfrac{-32}{3}$
   (c) $\dfrac{32}{3}$
   (d) $\dfrac{32}{9}$

5. Find the value of k in the equation $x^2 - 24kx + 23 = 0$

   (a) 0
   (b) 1
   (c) 2
   (d) 4

6. The equation $\left[25x^2 - \dfrac{40x}{3} + m\right] = 0$ has equal roots find m.

   (a) $\dfrac{9}{16}$
   (b) $\dfrac{4}{9}$
   (c) $\dfrac{9}{4}$
   (d) $\dfrac{16}{9}$

7. Solve $4x^2 - 24x + 35 = 0$

   (a) $\dfrac{3}{2}, \dfrac{5}{2}$
   (b) $\dfrac{5}{2}, \dfrac{7}{2}$
   (c) $\dfrac{9}{2}, \dfrac{3}{2}$
   (d) $\dfrac{9}{2}, \dfrac{11}{2}$

8. Solve $3x^2 + x - 24 = 0; (x > 0)$

   (a) $\dfrac{2}{3}$ (b) $\dfrac{4}{3}$ (c) 2   (d) $\dfrac{8}{3}$

9. If $ax^2 + bx + c = 0$ has equal roots; find c

   (a) $\dfrac{b}{4a}$ (b) $\dfrac{b^2}{a}$ (c) $\dfrac{b^2}{2a}$ (d) $\dfrac{b^2}{4a}$

10. If the sum of the roots of a quadratic equation $\alpha + \beta = 21$ and their difference $(\alpha - \beta)$ is 7, write the Q equation.

    (a) $x^2 - 21x - 98 = 0$ (c) $x^2 + 21x - 98 = 0$
    (b) $x^2 - 21x + 98 = 0$ (d) $x^2 + 21x + 98 = 0$

11. If for the quadratic equation $ax^2 + bx + c = 0$, one root is the reciprocal of the other, write the equation.

    (a) $ax^2 + bx - a = 0$ (c) $ax^2 + bx + a = 0$
    (b) $ax^2 - ax - b = 0$ (d) $ax^2 - ax + b = 0$

Solutions

1. The solutions of (i) are α, β satisfying

   $\alpha + \beta = b; \; \alpha\beta = c$, Then $(\alpha - \beta) = \sqrt{(\alpha+\beta)^2 - 4\alpha\beta} = \sqrt{b^2 - 4c}$
   and $\alpha^2 + \beta^2 = (\alpha+\beta)^2 - 2\alpha\beta = b^2 - 2c$
   $\therefore (\alpha^4 - \beta^4) = (\alpha+\beta)(\alpha-\beta)(\alpha^2+\beta^2)$
   $= b\sqrt{b^2 - 4c}\,(b^2 - 2c)$

2. $9x^2 + 15x + 4 = 0$

   $\Rightarrow 3x(3x+4) + 1(3x+4) = 0$
   $\Rightarrow (3x+4)(3x+1) = 0$
   $\therefore x = \dfrac{-4}{3}$ and $\dfrac{-1}{3}$

3. If $3+\sqrt{5}$ is a root of the equation then $3-i\sqrt{5}$ also will be a root, since the quantities m, n are real. Thus the two roots are $\alpha = 3+i\sqrt{5}$ and $\beta = 3-i\sqrt{5}$ and since $\alpha+\beta = -m$ and $\alpha\beta = n$, we have $(3+i\sqrt{5}+3-i\sqrt{5}) = 6 = -m$ and $(3+i\sqrt{5})(3-i\sqrt{5}) = 9+5 = 14 = n$
Thus the equation will be $x^2 - 6x + 14 = 0$ and the roots will be

$$\frac{6\pm\sqrt{36-56}}{2} = \frac{6\pm i\sqrt{20}}{2} \text{ or } 3\pm i\sqrt{5}$$

4. We have $\alpha+\beta = \frac{8}{3}$; and $\alpha\beta = \frac{4}{3}$

$\therefore \alpha^2 - \beta^2 = \frac{8}{3}(\alpha-\beta)$. Now $(\alpha-\beta) = \sqrt{\frac{64}{9} - 4 \cdot \frac{4}{3}} = \sqrt{\frac{64-48}{9}} = \frac{4}{3}$

$\therefore \alpha^2 - \beta^2 = \frac{8}{3} \cdot \frac{4}{3} = \frac{32}{9}$

5. The given equation is

$x^2 - 23kx - kx + 23 = 0$ since α, β are 23 and 1; $\alpha+\beta = 24k$; $\alpha\beta = 23$

Since α is prime; $\alpha = 23$ and $\beta = 1$

$\therefore \alpha+\beta = 24 = 24k$

$\therefore k = 1$

6. If α, β are the solutions; $\alpha+\beta = \frac{40}{3 \times 25} = \frac{8}{15}$ and $\alpha\beta = \frac{m}{25}$.

Since the roots are equal $\alpha+\beta = 2\alpha = \frac{8}{15}$

$\therefore \alpha = \frac{4}{15}$; $\therefore \alpha^2 = \frac{16}{225} = \frac{m}{25}$; $\therefore m = \frac{16}{225} \times 25 = \frac{16}{9}$

7. $\Rightarrow 4x^2 - 10x - 14x + 35 = 0$

$\Rightarrow 2x(2x-5) - 7(2x-5) = 0$

$\Rightarrow (2x-7)(2x-5) = 0$

$\therefore x = \frac{5}{2} \text{ or } \frac{7}{2}$

8. Given $= 3x^2 + 9x - 8x - 24 = 0$

$\Rightarrow 3x(x+3) - 8(x+3) = 0 \quad \Rightarrow (3x-8)(x+3) = 0$

$\therefore x = -3$ or $\dfrac{8}{3}$

$x = \dfrac{8}{3}$ is valid (since x is given + ve)

9. $\alpha + \beta = \dfrac{-b}{a}$; $\alpha\beta = \dfrac{c}{a}$. Since the roots are equal $2\alpha = \dfrac{-b}{a} \Rightarrow \alpha = -\dfrac{b}{2a}$

And $\alpha\beta = \alpha^2 = \dfrac{c}{a}$ $\therefore \alpha^2 = \dfrac{b^2}{4a^2} = \dfrac{c}{a}$ Or $c = \left(\dfrac{b^2}{4a}\right)$

10. $\alpha + \beta = 21$
$\alpha - \beta = 7$
Add $2\beta = 14$

$\therefore \alpha = 14$ (i)

Subtract $2\beta = 14 \qquad \beta = 7$ (ii)

The required equation will be $(x - \alpha)(x - \beta) = 0$

$\Rightarrow (x - 14)(x - 7) = 0 \quad \Rightarrow x^2 - 21x + 98 = 0$

11. If α and β are the roots and given $\beta = \dfrac{1}{\alpha}$ then $\alpha + \beta = \dfrac{-b}{a}$ i.e

$\left(\alpha + \dfrac{1}{\alpha}\right) = \left(\dfrac{-b}{a}\right)$ and $\alpha\beta = \alpha \cdot \dfrac{1}{\alpha} = 1 = \dfrac{c}{a}$ $\therefore c = a$

Also $\dfrac{\alpha^2 + 1}{\alpha} = \dfrac{-b}{a}$ (i)

Continuing further:

The equation is $ax^2 + bx + a = 0 \Rightarrow a(x^2 + 1) = -bx$

$\Rightarrow x^2 + \dfrac{b}{a}x + 1 = 0 \qquad \therefore x = \left[\dfrac{\dfrac{-b}{a} \pm \sqrt{\dfrac{b^2}{a^2} - 4}}{2}\right]$

The roots are $\dfrac{\dfrac{-b}{a} + \sqrt{\dfrac{b^2 - 4a^2}{a^2}}}{2}$ and $\dfrac{\dfrac{-b}{a} - \sqrt{\dfrac{b^2 - 4a^2}{a^2}}}{2}$

If these are α and $\dfrac{1}{\alpha}$ then

$$\alpha = \dfrac{\frac{-b}{a} + \frac{\sqrt{b^2-4a^2}}{a}}{2} \text{ and } \dfrac{1}{\alpha} = \dfrac{-b-\sqrt{b^2-4a^2}}{2a}$$

Then $\dfrac{-b+\sqrt{b^2-4a^2}}{2a} = \dfrac{2a}{-b-\sqrt{b^2-4a^2}}$

Crossing multiplying $b^2 - (b^2 - 4a^2) = 4a^2$

This is (trivial) an identity

Thus using (i) the equation given will be $ax^2 + bx + a = 0$

A. Linear Equations

| 1) | c | 2) | b | 3) | b | 4) | b |

B. Quadratic Equation

1)	c	2)	a	3)	c	4)	d	5)	b	6)	d
7)	b	8)	d	9)	d	10)	b	11)	c		

25

Alligation (Mixtures)

Mixtures and Adulteration

If two varieties with different cp's are mixed

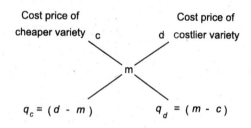

Fig. 1

Ratio in which the two varieties are to be mixed to get mean rate of m for a required profit for ex. $= \dfrac{q_c}{q_d} = \dfrac{(d-m)}{(m-c)}$ ...(i)

This is shown in fig 1

Exercises

1. If type A of rice costing Rs 18per kg and type B ratio costing Rs21/- per kg are mixed per kg are mixed in the ratio 4:5, what is the price per kg of the mixture?

   (a) Rs.19.37/-   (c) Rs.20.25/-   (e) Rs.21.75/-
   (b) Rs.19.67/-   (d) Rs.20.50/-

Alligation (Mixtures)

2. An unscrupulous ration shop dealer mixes 100kg of Rs2/- per kg with 50kg of Rice @ Rs.10/- per kg (type B) and sells the mixture for Rs.750/- what is the profit? What is the (CP of the mixture?)

   (a) Rs.42/-; Rs.5.33per kg (d) Rs.60/-; Rs.5.33per kg
   (b) Rs.45/-; Rs.5perkg   (e) Rs.65/-; Rs.5.67 per kg
   (c) Rs.50/-; Rs.4.67 per kg

3. 2kg of tea of type A costing Rs.150/- per kg, 2kg of type B costing Rs.175/- and 3kg of type c are mixed. The mixture is worth Rs.200per kg. What is the unit price of type c?

   (a) Rs.244/- (b) Rs.250/- (c) Rs.275/- (d) Rs.289/- (e) Rs.292/-

4. A shopkeeper has 800kg of rice; he sold a part to 10%profit, and the remaining part to a profit of 15%. On the whole he earned a profit of 12%, what amount is the first part?

   (a) 198kg (b) 285kg (c) 320kg (d) 400kg (e) 480kg

5. In what ratio should a grocer mix two varieties of rice costing Rs.1950/- per quintal and Rs.1450/- per quintal, so as to have a mixture of worth Rs.1850/- per quintal?

   (a) 1:1 (b) 1:2 (c) 1:3 (d) 1:4 (e) 1:5

6. Two liquids x and y are contained in a cistern in the ratio 3:2. 6 l of the mixture is taken out and the container is filled with 6 l of y. Now the ratio of x:y is 3:4. What is the volume of liquid y originally?

   (a) $\frac{42}{5}l$ (b) $\frac{49}{5}l$ (c) $\frac{52}{5}l$ (d) $\frac{57}{5}l$ (e) $\frac{63}{5}l$

7. Milk and water in a vessel x are in the ratio 5:4. Milk and water in the vessel y are in the ratio 3:4. In what ratio the mixtures are to be mixed to get a new mixture in vessel z containing half milk and half water?

   (a) 5:9 (b) 2:3 (c) 7:9 (d) 1:1 (e) 9:7

Quantitative Aptitude

8. In what ratio must water costing Rs2 per l be mixed with milk costing Rs.16/- per litre to get a mixture costing Rs.12/- per litre?
   (a) 1:4   (b) 2:5   (c) 3:5   (d) 5:3   (e) 5:2

9. Type I wheat costing Rs.12/- per kg is mixed with Type II wheat in the ratio 5:4. The mixture is priced at Rs.14/- per kg. What is the price of Type II wheat?

   (a) Rs.13.95/- per kg   (c) Rs.15.75/- per kg   (e) Rs.16.5/- per kg?
   (b) Rs.14.85/- per kg   (d) Rs.16/- per kg

10. Type A tea costing Rs.200/- per kg and type B costing Rs.110/- per kg are to be mixed to gain 20% profit by selling the mixture @ Rs.180/- per kg. In which proportion should A and B be adulterated?

    (a) 1:5   (b) 2:5   (c) 3:5   (d) 4:5   (e) 5:4

11. A milk vendor purchases milk from a dairy @ Rs.20/- per l. He wants to sells 50 l of mixture of milk and water @ Rs.12/- per l. How much milk should be purchased?

    (a) 10 l (b) 20 l (c) 30 l (d) 40 l (e) 50 l

Solutions

1. 4kg of type $A = 4 \times 18 = 72$

   5kg of type $B = 5 \times 21 = 105$

   Cost of 9kg of mixture $= 177$

   Rate $= \dfrac{177}{9} = $ Rs.19.67/- per kg

2. C.P. of type A rice = Rs.2/- × 100kg = Rs.200/-

   C.P. of type B rice = Rs.10/- × 50kg = Rs.500/-

   C.P. of 150kg mixture = Rs.700/-; S.P = Rs.750/-

   Gain = **Rs.50/-** ans

   C.P. of mixture $= \dfrac{Rs.700/-}{150 kg} = $ Rs 4.67per kg ans

Alligation (Mixtures) 193

3. 2kg Type A = $2 \times 150 \frac{Rs}{kg}$ = Rs.300/-

   2kg Type B = $2 \times 175 \frac{Rs}{kg}$ = Rs.350/-

   3kg Type C = $3 \times x \frac{Rs}{kg} = \frac{Rs\,3x}{Rs(650+x)}$ = Rs.7 × 200(given)

   7kg of mixture

   ∴ $3x = 1400 - 650 = $ Rs.750/-

   ∴ $x = $ Rs.250/kg

4. Let the first part be xkg;

   ∴ second = (800 - x kg)

   Total profit = $\frac{10}{100}x + \frac{15}{100}(800 - x) = \frac{12}{100} \times 800$ (given)

   ⇒ $12000 - 5x = 9600$

   ∴ $5x = 2400$

   ∴ $x = 480$ kg

5. Costlier rice, d = Rs.1950/- per quintal; cheaper rice c = Rs.1450/- per quintal; Mean value m Rs.1850/-

   The two varieties are to be mixed

   In the ratio $q_C : q_A = (d - m) : (m - c)$

   $= (1950 - 1850) : (1850 - 1450) = 100 : 400$

   Or 1:4 ratio

   ```
        c              d
         \            /
           1850 (m)
         /            \
       (d – m)        (m – c)
   ```

   Check: LHS $1450 \times 1 \times 1950 \times 4 = 9250$; RHS $1850 \times 5 = 9250$
   QED

6. Let q be the volume in l originally of the mixture. Quantity of X is $\frac{3}{5}q$ and Y is $\frac{2}{5}q$ in 1.6 l of mixture is taken off. It contains $6 \times \frac{3}{5}l$ of x and $6 \times \frac{2}{5}l$ of y. It is filled with 6 l of liquid y. Then the change in the quantities of x and y are

$$x : \left(\frac{3}{5}q - 6 \times \frac{3}{5}\right)l; \quad y : \left(\frac{2}{5}q - 6 \times \frac{2}{5} + 6\right)l$$

i.e. $\frac{3}{5}(q-6)l$ of x and $\frac{2}{5}q - 6 \times \frac{2}{5} + 6 \times \frac{5}{5}$ or $\frac{2}{5}q + \frac{18}{5}$ or $\frac{2}{5}(q+9)l$ of y.

The ratio after this operation of x and y is $\frac{3}{5}(q-6); \frac{2}{5}(q+9)$

This is given as 3:4

Cross multiplying and simplifying we get $q = 21 l$

Originally the vol. of x is $21 \times \frac{3}{5}$ or $\frac{63}{5}l$ ans

Check: original vol. of mixture $21\ l$ with $\frac{63}{5}l$ if x and $\frac{42}{5}$ of y and $\frac{6 \times 5}{5}$ of y

Now Qty of x is $\frac{63}{5} - \frac{18}{5} = \frac{45}{5}$ or $9l$;

y is $\frac{42}{5} - \frac{12}{5} + \frac{30}{5} = \frac{60}{5} = 12l$

Total is $21\ l$

$6l$ mixture is taken out i.e $\frac{6 \times 3}{5}$ of x and $\frac{6 \times 2}{5}$ of y; and $\frac{6 \times 5}{5}$ of y put in

7. Let cost price of milk be Rs.1/- per litre Milk in vessel

   $X = \frac{5}{9}$ CP of $1\ l$ mixture in vessel X is $\frac{5}{9}$ Rs.

   Milk in $1\ l$ mixture of vessel $y = \frac{3}{7}$ Cheaper component   C of 1 l of dearer component

   CP of $1\ l$ mixture in $y = \frac{3}{7}$

   Milk in $1\ l$ mixture in $Z = \frac{1}{2}l$

   CP of $1\ l$ mixture in $Z = \frac{1}{2}$

   The costlier component out of $\frac{5}{9}$ and $\frac{3}{7}$ is $\frac{35,27}{63} \rightarrow \frac{5}{9}$ i.e mixture x is cost litre

       c
        \
         Mean price
           (m)
        /   \
   (d − m)   (m − c)

Cost of 1 l from $X = \dfrac{5}{9} = d$ say

The cheaper component is $\dfrac{3}{7}$, its cost (1 l from y) is c say $= \dfrac{3}{7}$

Mean: 1 l of $Z = \dfrac{1}{2}$; $(d-m) = \dfrac{5}{9} - \dfrac{1}{2} = \dfrac{1}{18}$ and $(m-c) = \dfrac{1}{2} - \dfrac{3}{7} = \dfrac{1}{14}$

From the rule of alligation

$$\dfrac{\text{Qty of cheaper variety}}{\text{Qty of dearer variety}} = \dfrac{(d-m)}{(m-c)}$$

$$= \dfrac{\left(\dfrac{1}{18}\right)}{\left(\dfrac{1}{14}\right)} = \dfrac{7}{9} \text{ ans}$$

Thus the mixtures of vessel × (dearer and vessel)

Y (cheaper) are to be combined in the ratio 9:7

i.e 9 l of mixture from vessel X is be mixed with 7 l of mixture from Y to get 16 l of mixture in the vessel Z.

Check: Milk content in Z containing 16l of mixture $= 9 \times \dfrac{5}{9} + 7 \times \dfrac{3}{7} = 8l$

Water content in Z containing 16 l of mixture $= 9 \times \dfrac{4}{9} + 7 \times \dfrac{4}{7} = 8l$

Thus milk and water are same in quantities in vessel Z

8. Let w litres of water and m litres of milk be mixed.

Cost of water = Rs.2w/-; cost of milk = Rs.16m/-

Cost of (w+m) litres of mixture = Rs.12/- per l

$\therefore 2w + 16m = (w+m)12$

$\therefore \dfrac{w}{m} = \dfrac{2}{5}$ ans

9. $\left(5kg \times 12\dfrac{Rs}{kg}\right) + \left(4kg \times \dfrac{xRs}{kg}\right) = (5+4)kg \times 14\dfrac{Rs}{kg}$

$\therefore 60 + 4x = 126 \qquad \therefore x = \dfrac{66}{4} = 16.5\dfrac{Rs}{kg}$

10. SP of mixture = Rs.180/- per kg; profit = 20%. Hence CP of the mixture

$$\frac{100}{120} \times 180 = Rs\,150/kg$$

CP of one kg of dearer variety A
d = Rs. 200 per kg.

CP of one kg of cheaper variety B
C = Rs.10 per kg

Mean m = Rs.150/- / Kg.

(m−c) = 40   (d−m) = 50

$$\frac{\text{Qty of cheaper variety B}}{\text{Qty of dearer variety A}} = \frac{d-m}{m-c} = \frac{50}{40}$$

Thus A and B are to be mixed in the ratio of 4 to 5 ans

Check: (Type A = 4kg 200 × 4 + Type B = 5kg 110 × 5)
= (Mixture of 9kg) = 800 + 550 = 1350

Mean price = $\frac{1350}{9}$ = Rs.150/- per kg

11. CP of dearer component (milk) per l, c of cheaper component water per l

$$\frac{\text{Qty of milk}}{\text{Qty of water}} = \frac{m-c}{d-m} = \frac{12}{8} = \frac{3}{2}$$

For 50 l of mixture he has to procure 30 l of milk and 20 l of water.

CP of dearer component (milk)
per/ d = 20 Rs./ l

C of cheaper component (water)
per/ 0 Rs./ l

Mean cost price Rs. 12/-

(m−c) = 12   (d−m) = 8

Key

1)	b	2)	c	3)	b	4)	e	5)	d	6)	e
7)	c	8)	b	9)	e	10)	d	11)	c		

26

Simple Interest (S.I)

Exercises

Banking Problems

1. What is the simple interest rate percent per annum (pc pa) if a sum of money gets doubled in 10 years?

   (a) 7.25%   (b) 8.5%   (c) 10%   (d) 12%   (e) 13%

2. A certain amount given on S.I yields an amount of Rs.2500/- at the end of 4years, and Rs.3000/- at the end of 8y. What is the rate pc pa?

   (a) 6.25%   (b) 7.50%   (c) 8.75%   (d) 9.33%   (e) 10.25%

3. The S.I on a certain amount for 2years @ 8 pc pa rate is Rs.1400/- less than when lend for 3years @ 10 pc pa rate. What is the amount?

   (a) Rs.3000/-   (c) Rs.9000/-   (e) Rs.12,000/-
   (b) Rs.5000/-   (d) Rs.10,000/-

4. The interest (S.I) paid by a person in 10y on a loan taken @ 10 pc pa is Rs.60,000/- What is the amount borrowed?

   (a) Rs.52000/-   (c) Rs.56,000/-   (e) Rs.60,000/-
   (b) Rs.54000/-   (d) Rs.58,000/-

5. The S.I accrued on a certain amount in a housing loan scheme for 10y @ 4% rate is Rs.3,888/-. What is the amount borrowed by the employee?

   (a) Rs.9400/-   (c) Rs.9650/-   (e) Rs.9950/-
   (b) Rs.9600/-   (d) Rs.9720/-

6. The S.I on Rs.5000/- @ 5% rate was Rs.1250/-. What is the period in which the amount is repaid?

   (a) 1 y   (b) 2.5 y   (c) 3.2 y   (d) 4.6 y   (e) 5 y

7. Rs.5000/- becomes with S.I in 4years is Rs.6400/-. If the rate of interest is increased by 2%, what will be the amount in 4years?

   (a) Rs.6586/-   (c) Rs.6636/-   (e) Rs.6800/-
   (b) Rs.6606/-   (d) Rs.6750/-

8. A sum of Rs.10,000/- amounts to Rs.14000/- in 5years @ r pc pa on S.I. What is 'r'?

   (a) 6.5 pc pa   (b) 7.25 pc pa   (c) 8 pc pa
   (d) 9 pc pa   (d) 9.5 pc pa

9. X borrowed Rs.6400/-with a S.I for as many years as the rate of interest (numerically). He paid Rs.11,584/- at the end of the period. What is r?

   (a) 5%   (b) 6%   (c) 7%   (d) 8%   (e) 9%

10. A certain amount is given on S.I @ 8pc pa for 5years. An equal amount is given on S.I @10% for 8 y. What is the ratio of the interest?

    (a) 1:2   (b) 1:3   (c) 2:3   (d) 3:4   (e) 5:6

11. The S.I on a certain amount p for x years @ x pc pa is $100x^2$. What is p?

(a) Rs.8762/- (c) Rs.9000/- (e) Rs.11,000/-
(b) Rs.8850/- (d) Rs.10,000/-

12. Y borrowed Rs.1000/- on S.I from x @ 10% for 3y. z borrowed from y some amount (> 1000) @ 12 pc pa on S.I for 3y. In this deal y made a profit of Rs.420/- What is the additional amount (together with which y gave the Rs.1000/- he borrowed from x) y lend to z?

(a) Rs.1000/- (c) Rs.2000/- (e) Rs.3600/-
(b) Rs.1500/- (d) Rs.2500/-

13. In how much time a sum given on S.I @ 9pc pa yields an interest equal to 60% of the sum?

(a) 5y, 4m (b) 6y, 8m (c) 7y, 2m (d) 7y, 8m (e) 7y, 11m

14. In how much time a sum given on S.I @ 12 pc pa gives an interest of one fourth of the sum?

(a) 2y (b) 2y, 1m (c) 2y, 5m (d) 2y, 7m (e) 2y, 10m

Solutions

1. Let p be the principle, Amount = P + I, Given A = 2p
   $\Rightarrow$ P = I; n = 10y; r = ?

   $S.I = I = \dfrac{pnr}{100}$ = (given as) = P or $\dfrac{10r}{100} = 1 \therefore r = 10\%$

2. Let p be the principle; Amount $A_1 = 2500 = p + I_1$ in 4y

   $= p + \dfrac{p \times r 4}{100} \therefore 2500 - p = \dfrac{4pr}{100}$; \hfill (i)

   and after 8y; Amount $A_2 = 3000 = p + I_2 = p + \dfrac{p8r}{100}$

$\Rightarrow 3000 - p = \dfrac{8pr}{100}$ (ii) Subtracting (i) from (ii)

We have $500 = \dfrac{4pr}{100}$ (iii) $\therefore p = \dfrac{5,0000}{4r}$ (iv)

From (i) and (iii) $\dfrac{4pr}{100} = 2500 - p = 500$ $\therefore p = 2000$

and from (iii) $r = \dfrac{500 \times 100}{4 \times 2000} = \dfrac{500}{80} = \dfrac{50}{8} = \dfrac{25}{4} = 6\dfrac{1}{4}\%$

Check $\dfrac{pnr}{100} = 2000 \times 4 \times \dfrac{25}{4} \times \dfrac{1}{100}$

$= 500$ and $p + \dfrac{pnr}{100} = 2000 + 500 = 2500$ QED

3. If p is the principle; given $\dfrac{p \times 3 \times 10}{100} - \dfrac{p \times 2 \times 8}{100} = $ Rs.1400/-

$\therefore \dfrac{p}{100} 14 = 1400$

$\therefore p = \dfrac{1400 \times 100}{14} = 10,000$

4. Let p be the principle

Given S.I $= \dfrac{p \times 10 \times 10}{100} = 60,000$

P = Rs.60,000

5. Let p be the amount; Given $\dfrac{p \times 10 \times 4}{100} = 3888$

$\therefore P = \dfrac{3888 \times 100}{40} = \times 1944\dfrac{10}{2} = $ Rs.9720/-

6. P = Rs.5000/-; A = P + I $\therefore$ I = Rs.1250/- $= \dfrac{5000 \times n \times 5}{100}$

$\therefore n = \dfrac{1250 \times 100}{5000 \times 5} = 5$ years

7. Given $= p = 5000; r = r_1$ say; $n = 4y; A_1 = 6400;$

$A_1 = P + I$ where $I_1 = \dfrac{5000 \times 4 \times r_1}{100} = 200r_1$

Since $6400 = 5000 + 200r_1$ (given) ; $r_1 = \dfrac{1400}{200} = 7\%$

When $r_2 = 9\%$ then $I_2 = \dfrac{5000 \times 4 \times 9}{100} = 1800$

Then $A_2 = p + I_2 = 5000 + 1800 = $ Rs.6800/-

8. P = Rs.10,000/-; A = 14,000/-; n = 5y; r = ?

$I = (A - P) = 4000 = \dfrac{pnr}{100} = \dfrac{10,000 \times 5 \times r}{100} = 500r$

$\therefore r = \dfrac{4000}{500} = 8$ pc pa

9. P = Rs.6400/- ; n = r; A = Rs.11,584/-

$\therefore I = A - P$
$= 11584$
$- 6400$
$= 5184$

Now $I = 5184 = \dfrac{pnr}{100} = \dfrac{6400r^2}{100}$

$\therefore r^2 = \dfrac{5184 \times 100}{6400} = \dfrac{648}{8} = 81$

$\therefore r = 9\%$

$r = 9$ pc pa

10. Let p be the amount; $r_1 = 8$pc pa; $n_1 = 5y; r_2 = 10$pc pa; $n_2 = 8y$; Reqd $= I_1 : I_2$?

$I_1 = \dfrac{pn_1r_1}{100} = \dfrac{p \times 5 \times 8}{100}$; $I_2 = \dfrac{p \times 10 \times 8}{100}$

$\therefore \dfrac{I_1}{I_2} = \dfrac{5}{10} = \dfrac{1}{2}$

11. Let p be the principle. Given: $\dfrac{p \times x \times x}{100} = 100x^2$

$\therefore p = 100 \times 100 = $ Rs.10,000/-

12. $P_1 = 1000; n_1 = 3y; r_1 = 10$ pc pa;

$P_2 = (1000 + p_2^1); n_2 = 3y; r_2 = 12$ pc pa

$I_1 = \dfrac{p_1 n_1 r_1}{100} = \dfrac{1000 \times 3 \times 10}{100} = 300$

$I_2 = \dfrac{(1000 + p_2^1) 3 \times 12}{100}$ and given $I_2 - I_1 =$ Rs. 420/-

$\therefore (1000 + p_2^1)\dfrac{36}{100} - 300 = 420$

$\therefore (1000 + p_2^1)\dfrac{36}{100} = 720$

$\therefore (1000 + p_2^1) = \dfrac{720 \times 100}{36} = 1000$

$\therefore p_2^1 =$ Rs.1000/-

13. Let P be the sum; n = ? r = 9pc pa; $I = \dfrac{3}{5}p$

$I = \dfrac{3}{5}p = \dfrac{pn9}{100}$ $\therefore n = \dfrac{3p}{5} \dfrac{100}{9p} = \dfrac{20}{3} = 6\dfrac{2}{3}y$

i.e 6y 8m

14. Let p be the sum; $r = 12$ pc pa; $I = \dfrac{p}{4} n = ?$

$I = \dfrac{p}{4} = \dfrac{pn12}{100}$ $\therefore n = \dfrac{p}{4} \dfrac{100}{12p} = \dfrac{100}{48} = \dfrac{25}{12} = 2y \, 1m$

Key

1)	c	2)	d	3)	d	4)	e	5)	d	6)	e
7)	e	8)	c	9)	e	10)	a	11)	d	12)	a
13)	b	14)	b								

27

Compound Interest (C.I)

Exercises

1. Find the C.I on Rs. 8000/- @ 20 pc pa for 6 months compounded quarterly
   - (a) Rs. 420/-
   - (b) Rs. 535/-
   - (c) Rs. 625/-
   - (d) Rs. 715/-
   - (e) Rs. 820/-

2. A sum of Rs. 10,000/- each is given on loan for 2 year @ r pc pa to two parties. The difference between the C.I in one case and S.I in the other is Rs. 100/-. What is the rate pc pa?
   - (a) 8%
   - (b) 9%
   - (c) 10%
   - (d) 11%
   - (e) 12%

3. Find the C.I on Rs. 10,000/- @ r = 12 pc pa for 2y, 8 m compounded annually?
   - (a) Rs. 3198.12/-
   - (b) Rs. 3216.82/-
   - (c) Rs. 3381.32/-
   - (d) Rs. 3418.72/-
   - (e) Rs. 3547.52/-

4. Rs. 10,000/- gives an amount when given on C.I equal to Rs. 11881/- find the rate pc pa, n = 2 years; when compounded annually
   - (a) 8 pc pa
   - (b) 9 pc pa
   - (c) 10 pc pa
   - (d) 12 pc pa
   - (e) 15 pc pa

5. In how many years (n) will a sum of Rs. 10,000/- compound to 11248.64 at a C.I rate of 4 pc pa?
   - (a) 1 year
   - (b) 2 years
   - (c) 3 years
   - (d) 4 years
   - (e) 5 years

6. If the S.I on a sum of money @ 4 pc pa for 3 years is Rs. 1200/-. Find the C.I at the same rate for the same period.

   (a) Rs. 1205.82/-   (c) Rs. 1248.64/-   (e) Rs. 1272.10/-
   (b) Rs. 123.37/-   (d) Rs. 1289.30/-

7. A certain sum becomes Rs. 13,310/- in 3 y when given on C.I @ 10 pc pa. What will be the amount in 2 years, other things being same?

   (a) Rs. 11,900/-   (c) Rs. 13,000/-   (e) Rs. 13,600/-
   (b) Rs. 12,100/-   (d) Rs. 13,500/-

8. X has given a total loan of Rs. 21,000/- @ 10 pc pa on C.I he gives the first part to y, second part to z such that they amount to the same figure at the end of 2 and 3years respectively. What is p_2?

   (a) Rs. 10,000/-   (c) Rs. 12,000/-   (e) Rs. 14,000/-
   (b) Rs. 11,000/-   (d) Rs. 13,000/-

9. A sum of money amounts to Rs. 5,32,400/- in 3 years and Rs. 6,44,204 in 5y on C.I find the sum

   (a) Rs. 3,50,000/-   (c) Rs. 3,87,000/-   (e) Rs. 4,00,000/-
   (b) Rs. 3,75,000/-   (d) Rs. 3,95,000/-

10. Find the period in years if the C.I on Rs. 20,000/- is Rs. 3328/- @ 8 pc pa?

    (a) 1 y   (b) 11/2 y   (c) 2 y   (d) 21/2 y   (e) 3 y

11. Ramesh invested Rs. 6,000/- as fixed deposit on C.I @ 10 pc pa for 3y. How much amount will he get?

    (a) Rs. 7,000/-   (c) Rs. 7582/-   (e) Rs. 7986/-
    (b) Rs. 7318/-   (d) Rs. 7724/-

12. In how many years will a sum of Rs. 10,000/- @ 20 pc pa compounded semi annually become Rs. 13,310/-?

    (a) 1 y   (b) $1\frac{1}{2}$ y   (c) 2 y   (d) $2\frac{1}{2}$ y   (e) 3 y

13. Find the C.I on Rs. 10,000/- @ 16 pc pa in 3years

   (a) Rs. 5268.86/-   (c) Rs. 5487/-   (e) Rs. 5609/-
   (b) Rs. 5375.24/-   (d) Rs. 5593/-

14. Find the difference between the S.I and C.I on Rs. 1500/- n = 1y; r = 20 pc pa reckoned in the case of C.I; half yearly; in the case of SI the rate is same

   (a) Rs. 10/-   (c) Rs. 20/-   (e) Rs. 30/-
   (b) Rs. 15/-   (d) Rs. 25/-

15. Find the C.I on Rs. 80,000/- @ r = 20 pc pa compounded quarterly, for a 9 m period.

   (a) Rs. 12,580.95/-   (d) Rs. 12,600/-
   (b) Rs. 12,585.75/-   (e) Rs. 12,610/-
   (c) Rs. 12,596.64/-

Solutions

1. $P = 8000$; $r_a = 20$ pc pa; $n_p = \frac{1}{2}$ y; compounded quarterly

   $\therefore r = \frac{20}{4}$; $n_p = 2$ subscript p = period

   a = annual

   $\therefore A_c = p\left(1 + \frac{r}{4}\frac{1}{100}\right)^2 = 8000\left(1 + \frac{5}{100}\right)^2$

   $= 8000 \times \frac{21}{20} \times \frac{21}{20} = 8820$

   And $I_c = 8820 - 8000 =$ Rs. 820/-

2. (i) $P_1 =$ Rs. 10,000/-; $n_1 = 2y$; r_1 pc pa;

   $C.I = I_c = A_c - p = 10,000\left[1 + \frac{r}{100}\right]^2 - 10,000$;

   (ii) $P_2 = 10,000$; $n_2 = 2y$; $r_2 = r$ pc pa

   $SI = \frac{10,000 \times 2 \times r}{100}$

Difference = $10000\left[\left(1+\dfrac{r^2}{10^4}+\dfrac{2r}{100}-1\right)-\dfrac{2r}{100}\right]$

$= 10,000\left(\dfrac{r^2}{10^4}\right) = r^2 = 100$ (given)

∴ $r = 10$ pc pa

3. When interest is compounded annually, but time in fraction;

   $P = $ Rs. 10,000/-

   $r = 12$ pc pa;

   $n = 2\dfrac{8}{12} = 2\dfrac{2}{3} = \dfrac{8}{3}y$

   $A = 10,000\left[1+\dfrac{12}{100}\right]^2 \left[1+\left(\dfrac{2}{3}\times 12\right)\dfrac{1}{100}\right]$

   $= 10000\left(1+\dfrac{3}{25}\right)^2\left(1+\dfrac{2}{25}\right) = 13547.52$

   ∴ C.I = A − 1 = 13547.52 − 10,000 = Rs. 3547.52/−

4. $P = $ Rs. 10,000/−; C.I = 11881 − 10,000 = Rs. 1881/−; n = 2y

   Since $A = P\left(1+\dfrac{r}{100}\right)^2 = 11881$

   ∴ $\left(1+\dfrac{r}{100}\right)^2 = \dfrac{11881}{10000}$ ∴ $\left(1+\dfrac{r}{100}\right) = \dfrac{109}{100}$

   ∴ $100 + r = 109$ ∴ $r = 9$ pc pa

   $\begin{array}{r} 1\ \overline{1\ 8\ 8\ 1}\ |109 \\ 1 \\ \overline{209\ |\ 1881} \\ 1881 \\ \overline{} \end{array}$

5. $P = $ Rs. 10,000/−; $A = 11248.64$; $I = 1248.64$; $r = 4$pc pa; $n = ?$

   Since $A = 11248.64 = P\left(1+\dfrac{r}{100}\right)^n = 10,000\left(1+\dfrac{4}{100}\right)^n$

   ∴ $(1.04)^n = 1.124864$

   Trying values form $n = 1(1.04)^1 = 1.04$;

   $(1.04)^2 = 1.0816$; $(1.04)^3 = 1.124864$

Hence $n = 3$ (by trial and error)

Alternately trying various answer choice given; $n = 3$

6. Let p be the sum; $r = 4$ pc pa; $n = 3y$;

$$S.I = \frac{p \times 3 \times 4}{100} = 1200 \text{ (given)}$$

$$\therefore p = \frac{1200 \times 10^2}{12} = \text{Rs. } 10,000/-$$

$r = 4$ pc pa; $n = 3y$

$$C.I = 10,000 \left[1 + \frac{4}{100}\right]^3 - p = 10,000 \left(\frac{104}{100}\right)^3 - p$$

$= 11248.64 - 10000 = $ Rs. $1248.64/-$

7. Let p be the sum; $A = 13,310$; $n = 3y$; $r = 10$ pc pa

$$A = 13,310 = p\left(1 + \frac{10}{100}\right)^3$$

$$\therefore p = \frac{13310}{\left(\frac{11}{10}\right)^3} = \frac{13310 \times 10^3}{1331} = 10^4$$

Then $10^4 \left(1 + \frac{10}{100}\right)^2 = 10^4 \times \left(\frac{11}{10}\right)^2 = 10^2 \times 121 = 12,100$

8. Let p_1, p_2 be the parts; $p_1 + p_2 = $ Rs. $21,000/-$ — (i)

C.I; $r_1 = 10$ pc pa, $n_1 = 2y$;

$r_2 = 10$ pc pa, $n_2 = 3y$

$$A_1 = p_1 \left(1 + \frac{10}{100}\right)^2 ; A_2 = p_2 \left(1 + \frac{10}{100}\right)^3$$

Given that $A_1 = A_2$

$$\therefore p_1 \left(\frac{11}{10}\right)^2 = p_2 \left(\frac{11}{10}\right)^3 \Rightarrow p_1 = \frac{11p_2}{10} \quad \text{(ii)}$$

From (i) $\frac{11p_2}{10} + p_2 = \frac{21p_2}{10} = 21,000$

$\therefore p_2 = \dfrac{21000 \times 10}{21} = 10,000$ and $p_1 = 11,000$

9. Let p be the sum; r pc pa the rate; C.I

Given $p\left(1 + \dfrac{r}{100}\right)^3 = 5,32,400$ \hspace{1em} (i) and

$p\left(1 + \dfrac{r}{100}\right)^5 = 6,44,204$ \hspace{1em} (ii)

Dividing (ii) by (i)

$\left(1 + \dfrac{r}{100}\right)^2 = 1.21$

$\therefore 1 + \dfrac{r}{100} = 1.1$

$\therefore r = 0.1 \times 100 = 10$ pc pa

From (i) $p = \dfrac{5,32,400}{\left(1 + \dfrac{10}{100}\right)^3} = \dfrac{5,32400}{11^3} \times 10^3 = $ Rs. 4,00,000/-

10. P = 20,000; \hspace{1em} C.I = 3328; n = ?; \hspace{1em} r = 8 pc pa

C.I = $20000\left[\left(1 + \dfrac{8}{100}\right)^n - 1\right] = 3328$

$\Rightarrow 20000 \times \left(\dfrac{108}{100}\right)^n = 23328$

Trying different values n = 1, 2, 3... we find for n = 2 the equation is satisfied, \hspace{1em} n = 2y

(Alternately try the different answer choice and find n)

11. P = 6000; C.I ; r = 10 pc pa, n = 3y

$A = 6000\left(1 + \dfrac{10}{100}\right)^3 = 6000 \times \dfrac{1331}{1000} = $ Rs. 7986/-

12. P = 10,000, r = 20 pc pa = 10 pc pa, 6 m

n = ? \hspace{1em} A = Rs. 13310/-

$A = 13,310 = 10,000\left(1 + \dfrac{10}{100}\right)^{2n} = 10,000 \times \left(\dfrac{11}{10}\right)^{2n}$

This equation is satisfied when m = 3

$\therefore 10000 \times \dfrac{11}{10} \times \dfrac{11}{10} \times \dfrac{11}{10} = 13310$

$\therefore$ 2n = 3 or $n = 1\dfrac{1}{2}y$

13. P = Rs. 10,000/-; n = 3y; r = 16 pc pa; C.I = ?

$\text{C.I} = 10,000 \left[\left(1 + \dfrac{16}{100}\right)^3 - 1\right] = 10,000 \left[\dfrac{29 \times 29 \times 29}{25 \times 25 \times 25} - 1\right] \equiv$

or 10,000 (1.16³ - 1) = 10,000 (1.5609 - 1) = Rs. 5609/-

14. p = 1500; n = 1y; r = 20 pc pa

$\text{S.I} = \dfrac{1500 \times 1 \times 20}{100} = \text{Rs. 300/-}$

C.I = p = 1500; r_p = 10% (half yearly); n_p = 2 periods

$\text{C.I} = 1500 \left[\left(1 + \dfrac{10}{100}\right)^2 - 1\right] = 1500 \times \dfrac{121}{100} - 1$

$= \dfrac{1500 \times 21}{100} = 315$

Difference = 315 - 300 = Rs. 15/-

15. P = Rs. 80,000/-;   r = 20 pc pa compounded quarterly

$\therefore r_p = \dfrac{1}{4} \times 20 = 5pc$; n = 9m $\Rightarrow n_p = 3$ periods

$\text{C.I} = 80000 \left[\left(1 + \dfrac{5}{100}\right)^3 - 1\right] = 80000 \times \left[\dfrac{21 \times 21 \times 21}{20 \times 20 \times 20} - 1\right]$

$= 8 \times 10^4 \left[\dfrac{9261 - 8000}{8000}\right]$

$= 10 \times 1261 = \text{Rs. 12610/-}$

Key

1)	e	2)	c	3)	e	4)	b	5)	c	6)	c
7)	b	8)	b	9)	e	10)	c	11)	e	12)	b
13)	e	14)	b	15)	e						

28

Races and Skill Games

Defs: Terms

Dead heat race: If all the contestants (in a speed race) reach the goal simultaneously, the race is said to be a 'dead heat race'

If x is at the starting point of the race and y is ahead of 'a' m, we say x gives y a 'start' of 'a' m. y has to cover [S.D. i.e (stipulated distance) - a] while x has to cover the S.D.

In a 'r' m race (i.e when S.D = r m) if x covers 'r' m while y has to cover only (r - b)m, we say x 'beats' y by 'b' m or x gives y a 'start' of 'b' m.

Exercises

1. In a 2km race x beats y by 30m or 6sec. Find x's time over the course.

   (a) 4m 13s   (b) 5m 29s   (c) 5m 52s   (d) 6m5s   (e) 6m 34 s

2. If x covers the distance of 400m in 72 sec and y covers the same race in 80 sec, what is the distance by which x beats y?

   (a) 29m   (b) 32m   (c) 38m   (d) 40m   (e) 45m

3. P pedals a cycle 1.25 times as fast as Q. If p gives a start 60m, how far must be the winning point so that P and Q reach it simultaneously?

   (a) 270m   (b) 280m   (c) 300m   (d) 330m   (e) 350m

4. In a race of 200m P beats Q by 50m or 8sec. What time does P take to cover the course?

   (a) 20 sec   (b) 22 sec   (c) 24 sec   (d) 28 sec   (e) 32 sec

5. P beats Q by 100m or 48sec. In a 500m race what time does P take to cover the course?

   (a) 175 sec   (b) 192 sec   (c) 200 sec   (d) 220 sec   (e) 240sec

6. P runs a distance of 30m in the same time as Q covers 22.5m. In a km race Q beats P by

   (a) 230m   (b) 233.33m   (c) 240m   (d) 250m   (e) 260m

7. In a 500m race P has given a start of 25m to Q and to R a start of 37.5m. How many metres start can Q give R?

   (a) 13m   (b) 13.16m   (c) 13.64m   (d) 14.25m   (e) 15m

8. P's cycling speed is $\frac{4}{3}$ times that of Q. If P gives a start of 60m to Q; what is the length of the track if both reach the destination simultaneously?

   (a) 210m   (b) 220m   (c) 230m   (d) 240m   (e) 250m

9. In a 60m race P can beat Q by 10m, and Q can beat R by 6m. In a similar race P can beat R by

   (a) 10m   (b) 12m   (c) 15m   (d) 18m   (e) 20m

10. In a skill game, P can give Q 14 points out of 70, and R 21 points. How many points can Q give R in a game of 80 points?

    (a) 8 points   (c) 12 points   (e) 16 points
    (b) 10 points   (d) 14 points

11. If in a game of 60points P can give 4 points start to Q and 10 points to R how many can Q give R in a game of 112?

    (a) 6 points   (c) 10 points   (e) 14 points
    (b) 8 points   (d) 12 points

12. If P beats Q in a race of 90 points by 6 points, and R by 3 points, by how many points does R beat Q in a game of 174 points?

    (a) 6 points   (b) 7 points   (c) 8 points   (d) 11 points   (e) 13 points

13. In a 100m race P beats Q by 5m, and R by 8m. In a race of 190m, by how many metres does Q beat R?

    (a) 4m   (b) 6m   (c) 8m   (d) 9m   (e) 10m

Solutions

1. Y covers 30m in 6sec. ∴ y/s time over the course of 2km = $\frac{6 \text{ sec}}{30m} \times$ 2000m = 400 sec

    ∴ X's time over the course of 2km = (400 – 6) = 394 sec = 6m 34 sec

2. Y covered the distance in an extra 8 sec. X covers the distance 400m in 72 sec and Y in 80 sec. Their speeds are $\frac{400}{72} \frac{m}{s}$ and $\frac{400}{80} \frac{m}{\text{sec}}$ i.e $\frac{50}{9} \frac{m}{\text{sec}}$ and $\frac{5m}{\text{sec}} \equiv \frac{45m}{9 \text{ sec}}$

    Thus in 9 sec extra distance covered by x is 5m

    In 72 sec X covers 400m; and Y only $\frac{400}{80} \times 72 = 360m$.

    Thus X beats Y by 40m in this 400m race.

3. Speed of $P = S_P = \frac{5}{4} S_Q$. Let d be the dist.

    Times to reach $t_1 = t_2$

    $$\therefore \frac{d}{s_P} = \frac{d-60}{s_Q} \Rightarrow \left[\frac{d}{\left(\frac{5}{4} s_Q\right)}\right] = \left[\frac{d-60}{s_Q}\right]$$

    $$\therefore \frac{4}{5} \frac{d}{S} = \frac{d-60}{S}$$

    ∴ $4d = 5d - 300$ ∴ $d = 300m$

4. Q covers 50m in 8sec

∴ Q takes $\dfrac{8}{50m} \times 200m = 32$ sec to cover the course of 200m and

P covers in 32 - 8 = 24 sec

5. Q covers 100m in 48sec

∴ Q covers 500m in $\dfrac{48\,\text{sec}}{100m} \times 500m = 240$ sec

P covers 500m in 240 - 48 = 192 sec ans

6. Q covers a distance of 30m in the same time as p run in 22.5 m

∴ Q covers 1000m distance, while p runs

$\dfrac{22.5}{30} \times 1000 = 750m$

Thus in a 1km race Q beats P by 1000 - 750 = 250m

7. P covers 500m during the time Q covered (500-25) = 475m and during the time when R covers (500 - 37.5) = 462.5m. Thus Q covers 475m in the same time as R covers 462.5m. In the time when Q covers 500m, R covers $\dfrac{462.5}{475} \cdot 500 = 486.842m$

i.e Q can give (500 - 486.84) = **13.16m** start to R ans

8. $S_P = \dfrac{4}{3} S_Q$; if CB = x then $\dfrac{x+60}{S_P} = \dfrac{x}{S_Q} \Rightarrow \left(\dfrac{x+60}{\frac{4}{3}S_Q}\right) = \left(\dfrac{x}{S_Q}\right)$

$\Rightarrow x = 180$; Thus the length of the track = 180 + 60 = 240m

Check: $\dfrac{240}{S_P} hr = \dfrac{18}{S_Q} hr$ QED

9. $\dfrac{P}{Q} = \dfrac{60}{50}; \dfrac{Q}{R} = \dfrac{60}{54}$. Then $\dfrac{P}{R} = \dfrac{P}{Q} \cdot \dfrac{Q}{R} = \dfrac{60}{50} \cdot \dfrac{60}{54} = \dfrac{6}{5} \cdot \dfrac{10}{9} = \dfrac{12}{9}$

$= \dfrac{4}{3} \equiv \dfrac{60}{45}$

Thus P can beat R by 15m in a similar race of 60m

10. $\dfrac{P}{Q} = \dfrac{70}{70-14} = \dfrac{70}{56}$; $\dfrac{P}{R} = \dfrac{70}{70-21} = \dfrac{70}{49}$;

$\therefore \dfrac{Q}{R} = \dfrac{Q}{P} \cdot \dfrac{P}{R} = \dfrac{56}{70} \times \dfrac{70}{49} = \dfrac{8}{7} = \dfrac{80}{70}$

Q can give 10 points to R

11. $\dfrac{P}{Q} = \dfrac{60}{56}$; $\dfrac{P}{R} = \dfrac{60}{50}$;

$\therefore \dfrac{Q}{R} = \dfrac{Q}{P} \cdot \dfrac{P}{R} = \dfrac{56}{60} \times \dfrac{60}{50} = \dfrac{56}{50} = \dfrac{112}{100}$

Thus Q can give 12points to R in a game of 112 points

12. $\dfrac{R}{P} = \dfrac{87}{90}$; $\dfrac{P}{Q} = \dfrac{90}{84}$

$\therefore \dfrac{R}{Q} = \dfrac{R}{P} \cdot \dfrac{P}{Q} = \dfrac{87}{90} \cdot \dfrac{90}{84} = \dfrac{87}{84} = \dfrac{174}{168}$

R beats Q in 6points in a game of 174 points.

13. $\dfrac{P}{Q} = \dfrac{100}{95}$; $\dfrac{P}{R} = \dfrac{100}{92}$

$\therefore \dfrac{Q}{R} = \dfrac{Q}{p} \cdot \dfrac{P}{R} = \dfrac{95}{100} \cdot \dfrac{100}{92}$

$= \dfrac{95}{92} \equiv \dfrac{190}{184}$

Key								
1)	e	2)	d	3)	c	4)	c	5)	b	6)	d	7)	b
8)	d	9)	c	10)	b	11)	d	12)	a	13)	b		

29

Calendar

Procedure to find the day for the given date

Note: The codes mentioned in the steps 1 - 8 are given after step 8

Step 1 : Write the digits of the date

Step 2 : Write the month's code (Note that there is a slight variation for the code number if it is a leap year for the months of January and February)

Step 3: Write the last two digits of the year

Step 4 : Write the number of leap year elapsed: (by dividing the last two digits of the year by 4)

Step 5 : Write the century code

Step 6 : Add all the values in step 1 through step 5

Step 7 : Divide this total by seven; find the remainder

Step 8 : Read out the day corresponding to this remainder.

Codes to be used

(A) Code for centuries

Code for centuries	Code no	Leap year
16th century (1500 - 1599)	0	The four digit
17th century (1600 - 1699)	6	number should be
18th century (1700 - 1799)	4	divisible by 4
19th century (1800 - 1899)	2	(Last two digits to be
20th century (1900 - 1999)	0	divisible by 4)
21st century (2000 - 2099)	6	
22nd century (2100 - 2199)	4	

(B) Code for months

January	1	July	0	
February	4	August	3	
March	4	September	6	
April	0	October	1	
May	2	November	4	
June	5	December	6	

(C) Code for days

Saturday	0
Sunday	1
Monday	2
Tuesday	3
Wednesday	4
Thursday	5
Friday	6

Perpetual Calendar

Calendars for the years 1893 through 2032 are given in a consolidated form in this. The dates for any date is found from this in the following manner

Procedure: To use the perpetual calendar: Move around the table in the clock wise fashion starting at top left corner.

Step 1 : Read the (alphabetical) code for the given year (letters A up to N)

Step 2 : For the letter in the alphabet found in step1, read the number that falls under the particular given month (November from 1 to 7 given)

Step 3 : For the number read out in step 2 go to the particular date given in the column (bearing the number read out in)

Step 4 : And against the date read from the left hand list of day the corresponding day. Copy of the perpetual calendar is affixed.

Perpetual Calendar

Years				
1893 G	1928 N	1963 D	1998 D	
1894 A	1929 B	1964 J	1999 E	
1895 B	1930 C	1965 E	2000 M	
1896 J	1931 D	1966 F	2001 A	
1897 E	1932 L	1967 G	2002 B	
1898 F	1933 G	1968 H	2003 C	
1899 G	1934 A	1969 C	2004 K	
1900 A	1935 B	1970 D	2005 F	
1901 B	1936 J	1971 E	2006 G	
1902 C	1937 E	1972 M	2007 A	
1903 D	1938 F	1973 A	2008 I	
1904 L	1939 G	1974 B	2009 D	
1905 G	1940 H	1975 C	2010 E	
1906 A	1941 C	1976 K	2011 F	
1907 B	1942 D	1977 F	2012 N	
1908 J	1943 E	1978 G	2013 B	
1909 E	1944 M	1979 A	2014 C	
1910 F	1945 A	1980 I	2015 D	
1911 G	1946 B	1981 D	2016 L	
1912 H	1947 C	1982 E	2017 G	
1913 C	1948 K	1983 F	2018 A	
1914 D	1949 F	1984 N	2019 B	
1915 E	1950 G	1985 B	2020 J	
1916 M	1951 A	1986 C	2021 E	
1917 A	1952 I	1987 D	2022 F	
1918 B	1953 D	1988 L	2023 G	
1919 C	1954 E	1989 G	2024 H	
1920 K	1955 F	1990 A	2025 C	
1921 F	1956 N	1991 B	2026 D	
1922 G	1957 B	1992 J	2027 E	
1923 A	1958 C	1993 E	2028 M	
1924 I	1959 D	1994 F	2029 A	
1925 D	1960 L	1995 G	2030 B	
1926 E	1961 G	1996 H	2031 C	
1927 F	1962 A	1997 C	2032 K	

Months

	Jan	Feb	Mar	Apr	May	Jun	Jul	Aug	Sep	Oct	Nov	Dec
A	1	4	4	7	2	5	7	3	6	1	4	6
B	2	5	5	1	3	6	1	4	7	2	5	7
C	3	6	6	2	4	7	2	5	1	3	6	1
D	4	7	7	3	5	1	3	6	2	4	7	2
E	5	1	1	4	6	2	4	7	3	5	1	3
F	6	2	2	5	7	3	5	1	4	6	2	4
G	7	3	3	6	1	4	6	2	5	7	3	5
H	1	4	5	1	3	6	1	4	7	2	5	7
I	2	5	6	2	4	7	2	5	1	3	6	1
J	3	6	7	3	5	1	3	6	2	4	7	2
K	4	7	1	4	6	2	4	7	3	5	1	3
L	5	1	2	5	7	3	5	1	4	6	2	4
M	6	2	3	6	1	4	6	2	5	7	3	5
N	7	3	4	7	2	5	7	3	6	1	4	6

Calendars

	1	2	3	4	5	6	7
Mon	1						
Tue	2	1					
Wed	3	2	1				
Thu	4	3	2	1			
Fri	5	4	3	2	1		
Sat	6	5	4	3	2	1	
Sun	7	6	5	4	3	2	1
Mon	8	7	6	5	4	3	2
Tue	9	8	7	6	5	4	3
Wed	10	9	8	7	6	5	4
Thu	11	10	9	8	7	6	5
Fri	12	11	10	9	8	7	6
Sat	13	12	11	10	9	8	7
Sun	14	13	12	11	10	9	8
Mon	15	14	13	12	11	10	9
Tue	16	15	14	13	12	11	10
Wed	17	16	15	14	13	12	11
Thu	18	17	16	15	14	13	12
Fri	19	18	17	16	15	14	13
Sat	20	19	18	17	16	15	14
Sun	21	20	19	18	17	16	15
Mon	22	21	20	19	18	17	16
Tue	23	22	21	20	19	18	17
Wed	24	23	22	21	20	19	18
Thu	25	24	23	22	21	20	19
Fri	26	25	24	23	22	21	20
Sat	27	26	25	24	23	22	21
Sun	28	27	26	25	24	23	22
Mon	29	28	27	26	25	24	23
Tue	30	29	28	27	26	25	24
Wed	31	30	29	28	27	26	25
Thu		31	30	29	28	27	26
Fri			31	30	29	28	27
Sat				31	30	29	28
					31	30	29
						31	30
							31

These charts enable you to find the day of the week in any year from 1893 through 2032

Instructions

Step 1 : Find the year that you are interested and note the letter that follows it.

Step 2 : Find that letter above and note which number falls under the month you are looking for.

Step 3 : Use the calendar that bears the number you found in step 2

Calendar 219

Exercises

In the following examples, the procedure is illustrated for various conditions like (i) examples for 19th, 2012, 21st centuries (ii) for leap year and ordinary years (iii) for different months ranging from Jan, Feb, July, Aug, Nov; etc (iv) for dates before and after Feb 28th and 29th in code of leap year) and on 28th or 29th Feb.

1. Find the day of 26/01/2005

    (a) Monday   (c) Friday   (e) Sunday
    (b) Wednesday   (d) Saturday

2. Find the day of 15/8/2032

    (a) Thursday   (c) Saturday   (e) Monday
    (b) Friday   (d) Sunday

3. Find the day of 10/2/2032

    (a) Tuesday   (c) Thursday   (e) Saturday
    (b) Wednesday   (d) Friday

4. Find the day of 29/2/2024

    (a) Monday   (c) Wednesday   (e) Friday
    (b) Tuesday   (d) Thursday

5. 29/12/1999

    (a) Wednesday   (c) Friday   (e) Sunday
    (b) Thursday   (d) Saturday

Solutions

1. 26/01/2005   o.y

Digits of the date Twenty sixth	26
Month's code January	01 (o y)
Last two digits of the year (2005)	05
Leap years elapsed $\frac{05}{4} = 0/-1R$	01: $\frac{\text{(Last two digits of years)}}{4}$
Century code	06
Total	39

   Divide by 7: $\frac{7\lfloor 39}{5-4R}$. Remainder 4 corresponds to Wednesday.

   Checked with 2005 calendar tallies QED

2. 15/8/2032   Leap year

Digits of the date fifteenth	15
Month's code August	03 (though leap year)
Last two digits of the year (2032)	32
Leap years elapsed $\frac{32}{4}$	08: $\frac{\text{(Last two digits of year)}}{4}$
Century code 21st century (2000 - 2099)	06
Total	64

   Divide by 7: $\frac{7\lfloor 64}{9-1R}$. Remainder 1 corresponds to Sunday

   Checked with 'perpetual calendar' (Procedure and 'algorithm for codes' given after these solutions).

   Step 1 year code 'k'

   Step 2 Month code 'k' → 7

   Step 3 For month code for year's code '7' → for date 15th → Day is 'Sunday'

   Tallies with the solution mentioned 'Sunday' QED

3. 10/2/2032   Leap year

Digits of the date (Tenth)	10
Month's code (February)	03 (though leap year)
Last two digits of the year (2032)	32
Leap years elapsed $\frac{32}{4} = 08$	08: $\frac{\text{(Last two digits of years)}}{4}$
Century code 21st century (2000 - 2099)	06
Total	59

   Divide by 7: $\frac{7\lvert 59}{8 - 3R}$. Remainder 3 corresponds to Tuesday

   Checked with 'perpetual calendar

   Step 1 year 2032 corresponds to code k.

   Step 2 (code k for the year) corresponds for the month February to code 7

   Step 3 (Code 7 for the month) corresponds for the date tenth to Tuesday. This tallies

   Note: perpetual calendar available from my personal diary contains information up to 2032 year only (inclusive)

4. 29/2/2024 (leap year)

Digits of the date Twenty-ninth	29
Month's code (February)	03 (leap year)
Last two digits of the year (2024)	24
Leap years elapsed $\frac{24}{4} = 06$	06: $\frac{\text{(Last two digits of year)}}{4}$
Century code Twenty first (2000 - 2099)	06
Total	68

   Divide by 7: $\frac{7\lvert 68}{9/- 5R}$. Remainder 5 corresponds to Thursday

   Check: With perpetual calendar

Step 1 : year 2024 corresponds to code H.
Step 2 : Code H corresponds for the month February to 4
Step 3 : Code 4 for the month corresponds for the date 29th Thursday
This tallies QED

5. 29.12/1999   (o y)

Digits of the date (Twenty ninth)	29
Month's code (December)	06 (o y)
Year's last two digits	99
Leap years elapsed $\frac{99}{4}$	24: $\frac{\text{(Last two digits of year)}}{4}$
Century code 20th (1900 - 1999)	00
Total	158

Divide by 7 : $\frac{7\lfloor 158}{22/-4R}$. Remainder 4 corresponds to Wednesday

Check: Calendar 1999 tallies

Key

| 1) | b | 2) | d | 3) | a | 4) | d | 5) | a |

30

Clocks

Clocks

In 60 minutes the minutes pointer gains 55 minutes on the hours pointer (A)

- Both the pointers (minutes pointer and hours pointer) coincide only once in every hour
- When the two pointers are at right angles, they are 15 'minute spaces' apart, when in opposite directions, '30minute spaces' apart.
- Angle traced by the hour hand in 12 hours i.e in 12 × 60minutes is 360° i.e 1° in 2 minutes
- Angle traced by the minutes points in 60min = 360° i.e 6° in one minute
- When the correct time is 6'0clock and if a (wrong) clock indicates 6.05, it is said to be fast by 5minutes
- The pointers of a clock coincide 11 times in every 12 hours.
- The pointers coincide 22times in a day
- In 12 hours the pointers coincide **or** are in opposite directions **22** times.
- In 24 hours the pointers coincide **or** are in opposite directions **44** times.
- In 12 hours the pointers are at right angles **22** times
- In 24 hours the pointers are at right angles **44** times
- The pointers will be in opposite directions 11 times in every 12 hours because between 5 and 7 they point in opposite directions at 6'oclock only
- So in a day, the pointers point in opposite directions 22 times.
- In what follows unless otherwise mentioned all the [Angles are measured from North direction; In clock wise direction.]

Exercises

1. What is the angle between the hour pointer and minute pointer at 5.30 hours?

    (a) 15° (b) 20° (c) 25° (d) 32° (e) 34°

2. What is the angle between the pointers at 2.50 hours?

    (a) 200° (b) 207° (c) 209° (d) 215° (e) 220°

3. What is the angle between the hour pointer and the minute pointer at 1.30 am?

    (a) 119° (b) 120° (c) 130° (d) 135° (e) 140°

4. A watch is set correct at 6 am on 24th Jan. It gains 12 min in 24 hours. What will be the correct time it this (wrong) clock indicates 7 am on 29th Jan?

    (a) 6 am (b) 6.10 am (c) 6.40 am (d) 7.20 am (e) 7.40 am

5. The minutes pointer of a (wrong) clock overtakes the hours pointer at intervals of 66 min of the correct time. How much in a day the clock gains or loses?

    (a) loss 10.98 min (c) gains 5.90 min (e) gains 11.90 min
    (b) loss 11.90 min (d) gains 6.99 min

6. A watch gains 5 sec in 4 minutes. It was set right at 6 am. In the same day when the clock indicates 10.05 am, what is the correct time?

    (a) 9.45 am (c) 10 am (e) 10.20 am
    (b) 9.55 am (d) 10.15 am

7. At what time between 5'o clock and 6'o clock will the hands of a clock point in opposite directions?

    (a) 5 hr 41 min (c) 5 hr 55 min (e) 6.00 hr
    (b) 5 hr 49 min (d) 5 hr 58 min

8. A clock is set right at 6 am on 24 Jan. The clock loses 12 min in 24 hours. What will be the true time when the wrong clock indicates 5 am on 29 Jan?

   (a) 6 am   (c) 6.20 am   (e) 6.45 am
   (b) 6.04 am   (d) 6.40 am

9. A clock is started at 6 am. At 48 min past 10′o clock the hours pointer turned through how many degrees?

   (a) 140°   (b) 144°   (c) 150°   (d) 158°   (e) 172°

10. What is the angle between the two pointers at 9.20 hours?

    (a) 150°   (b) 155°   (c) 160°   (d) 165°   (e) 170°

11. How much time does a clock lose per day, if its pointers coincide every 63 minutes?

    (a) 0.852 hr   (b) 0.935 hr   (c) 1.012 hr   (d) 1.12 hr   (e) 1.15 hr

12. A watch gains uniformly; It was 3 m slow at 6 am on Monday the 2nd June, and fast by 5 min 36 sec on June 11th 5 am. When was it correct?

    (a) 7 am 9th June   (c) 9 am 5th June   (e) 9 am 6th June.
    (b) 8 pm 7th June   (d) 4 pm 5th June

13. A correct clock shows 6 am on 2nd June. In $1\frac{7}{8}$ days i.e. 45 hours how many degrees will the hour hand rotate through?

    (a) 1290°   (b) 1310°   (c) 1320°   (d) 1350°   (e) 1400°

Solutions

1. Angle traced by the hour pointer in 12 hr = 360° in 5 hr 30 min

   From N → $\frac{360}{12} \times \frac{11}{2} = 165°$

   Angle traced by the minute pointer in 1 hour = 360°

   In 30 minutes : $\frac{360}{60} \times 30 = 180°$

   Angle between the two pointers = 180 - 165 = 15°

2. At 2'o clock the angle from the north line ON is $\frac{2}{3} \times 90 = 60°$ (i)

In 12 hours the hour hand moves through 360°

∴ In 50 min → $\frac{360}{12 \times 60} \times 50 = 25°$ (ii)

So, at 2.50 hours the hour hand will be at $60 + 25 = 85°$ away from ON.

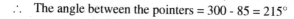

In one hour the minute hand moves through 360°

∴ In 50 min → $\frac{360}{60} \times 50 = 300°$

∴ The angle between the pointers = 300 - 85 = 215°

3.

At 1'o clock the angle between the two pointers is 30°

In 12 hours the hour hand moves through 360°

∴ In 30 min it moves through $\frac{360}{12 \times 60} \times 30 = 15°$

Thus the hour pointer will be $(30 + 15) = 45°$ away from the north line

In 60 min the minute hand covers 360°

In 30 min it covers 180°

Hence the angle between the pointers will be 180 - 45 = 135°

4. The time that elapsed between 6 am on 24th Jan and 7 am 29th Jan is $5 \times 24 + 1 = 121$ hours.

The 'wrong' clock gains 12 minutes in 24 hours ie when the wrong clock shows 24 hr 12 min i.e $24\frac{12}{60} = 24\frac{1}{5} = \frac{121}{5}$ hours. Actually, the time elapsed is 24 hours.

When the wrong clock shows 121 hours; actually it will be $\dfrac{24 \text{ hours}}{\left(\dfrac{121}{5} \text{ hours}\right)} \times 121$ hours = 120 hours i.e $\dfrac{120}{24} = 5$ days only. Thus the correct time will be 6 am only on 29, 12 Jan, not 7 am.

5. The minutes hand over takes the hour hand in 66 min. in this watch. In a correct clock the minutes hand gains 55 minute spaces over the hours pointer, in 60 minutes. To be together again, the minutes pointer must gain 60 'minute' spaces over the hours pointer. 55 minute spaces gained in 60minutes

∴ 60 minute spaces are gained in $\dfrac{60}{55} \times 60 = \dfrac{720}{11} = 65\dfrac{5}{11}$ minutes.

But they are together in the wrong clock after 66 min. The 'loss' in 66 minutes is $66 - 65\dfrac{5}{11} = \dfrac{6}{11}$ min

Hence, 'loss' in 24hours will be $\left(\dfrac{6\min}{11}\right) \times \dfrac{24 \times 60 \text{ minutes}}{66 \text{ minutes}}$

$= \dfrac{24 \times 60}{121} = 11.90$ minutes

∴ In a day it loses **11.9 minutes**

6. The wrong watch gains 5s in 4 min i.e it shows 245 sec when actually 240 sec elapses. When it shows the elapsing of (10.05 - 6) ie 4 hr 5 min i.e $4 \times 60 + 5 = 245$ min actually 240 min elapse. So, the correct time is 10 am only (c)

7.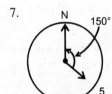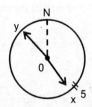

The angle between the pointer at 5'o clock is 90 + 60 = 150°. Let in x minutes they be together. In opposite directions:

In 12 × 60 min the hour pointer covers 360° min. hour.

In x min, then $\left(\dfrac{360°}{12 \times 60\,\text{min}} \times x\,\text{min}\right) = \left(\dfrac{x}{2}\right)°$

∴ The angle No y $= \left(150 + \dfrac{x}{2}\right)°$

Angle No y should be $150 + \dfrac{x}{2} + 180 = \left(330 + \dfrac{x}{2}\right)°$

In 60 min the minutes hand covers 360°

⇒ ie 360° is covered in 60 minutes by the minutes hand

∴ $\left(330 + \dfrac{x}{2}\right)°$ will be covered in $\left[\dfrac{60\,\text{min}}{360°} \times \left(330 + \dfrac{x}{2}\right)°\right]$

Thus x minutes $= \left[\dfrac{\left(330 + \dfrac{x}{2}\right)}{6}\right]$ min (Required result)

∴ $6x = 330 + \dfrac{x}{2}$ ⇒ $\dfrac{11}{2}x = 330$

∴ $x = 330 \times \dfrac{2}{11} = 60$ minutes

i.e at 6'o clock the pointers will be in opposite directions

8. The time elapsed between 6 am 24 Jan and 5 am 29th Jan is 5 × 24 − 1 = 119 hours. The clock loses 12 min in 24 hours. i.e 23 hr 48 min

i.e $23\dfrac{48}{60} = 23\dfrac{4}{5} = \dfrac{119}{5}$ hr in the wrong clock will be 24 hr in the correct clock

∴ 119 hr in the wrong clock will be $\dfrac{24\,hr}{\left(\dfrac{119}{5}\,hr\right)} \times (119\,hr)$

= 24 × 5 = 120 hr in the correct clock ie the correct time will be on 29th not 5 am, but 6 am

Clocks 229

9.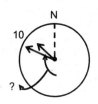

In 12 hrs the hr pointer moves through 360°

$$\therefore \text{In } 4\frac{48}{60} \text{ hours} = 4\frac{4}{5} \text{ hr}$$ it moves through $\left(\frac{360°}{12 \ hr} \times \frac{24}{5} hr\right) = 144°$

10.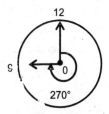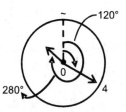

At 9'o clock the angle between the two pointers is 270°

In 12 hours the hour pointer covers 360°

$\therefore$ In 20 min it covers $\dfrac{360°}{12 \times 60 \ min} \times 20 \ min = 10°$

The hour hand will therefore now be 270° + 10° = 280° from ON.

In 60 min the minute hand covers 360°

In 20 min it covers $\dfrac{360°}{60 \ min} \times 20 \ min = 120°$ from ON

Hence the angle between the pointers = 280° - 120° = 160°

11. In the correct clock the minutes pointer gains 55 minute spaces over the hours pointer in 60 min.

To coincide the minutes pointer must gain 60 minutes spaces over the hours pointer $\left(\dfrac{60 \ minutes}{55 \ min \ spaces} \times 60 \ min. \ spaces\right) = \dfrac{720}{11} = 65\dfrac{5}{11}$ min

In the wrong clock they coincide in 63 minutes i.e the loss in 63 minutes is $65\dfrac{5}{11} - 63 = 2\dfrac{5}{11} = \dfrac{27}{11}$ min

Hence the loss in 24 hours will be $\dfrac{\frac{27}{11}\text{ min}}{63\text{ min}} \times (24 \times 60 \text{ min})$

$= \dfrac{27 \times 24 \times 60}{11 \times 63} = \dfrac{4320}{7 \times 11}$ minutes

$= \dfrac{4320}{7 \times 11 \times 60}$ hours $= \dfrac{72}{77} = 0.9351$ hours $\simeq 0.935\ hrs$

12. It gains (3 min + 5 min 36 sec) i.e 8 min 36 sec

    i.e 516 sec in (8 days & 23 hr) = 215 hr

    8 min 36 sec i.e $8\dfrac{36}{60} = 8\dfrac{3}{5}$ i.e $\dfrac{43}{5}$ min are gained in 215 hours.

    $\therefore$ 3 min are gained in $\dfrac{215\ hr}{\left(\dfrac{43}{5}\right)\text{min}} \times 3$ min

    $= \dfrac{215 \times 5 \times 3}{43} = 75$ hrs $= 3$ days 3 hours.

    It will be correct at 09 hr on 5th June i.e 9 am 5th June

13. In 12 hours hour hand rotates 360°.

    In 45 hours $\to \dfrac{360°}{12\ hr} \times 45\ hr = 1350° = \dfrac{1350}{360} = 3\dfrac{270}{360} = \left(\dfrac{15}{4}\right)$

    i.e. 3 round and 270°

Key

1)	a	2)	d	3)	d	4)	a	5)	b	6)	c
7)	e	8)	a	9)	b	10)	c	11)	b	12)	c
13)	d										

31

Stocks & Shares

Definitions and formulae

A group of people starting an industry may invite people to invest in the project, by assigning shares in the company.

Share certificates are issued to them 'dividends', (annual profits), promised. 'Face Value' is the value of a share or stock printed on the share certificate ('Normal Value' (NV or FV)).

Stocks of different companies are sold in the open market through brokers at stock exchanges at 'Market Value' (MV)

A share or stock is said to be

i) 'at premium' or above par if MV > FV

ii) 'at par' if MV = FV or (iii) at discount or below par if MV < FV

Broker charges some amount called 'brokerage'.

i) When stocks are purchased brokerage is added to the cost price (CP)

ii) When stock is sold brokerage is subtracted from selling price (SP)

'Dividend' is always paid on the FV

No. of share held by a person $= \left(\dfrac{\text{Total Investment}}{\text{Inv. in one share}}\right)$

$\left(\dfrac{\text{Total Income}}{\text{Income from one share}}\right) = \left(\dfrac{\text{Total FV}}{\text{FV of one share}}\right)$

Example: "A Rs.100; 8% stock at 110" means (i) FV (NV) of stock = RS.100/- (ii) MV = 110 (iii) Annual dividend on one share = 8% of FV = $\dfrac{8}{100}$ of 100 = Rs.8/- (iv) An investment of Rs.110/- gives an annual income of Rs.8/- (v) Rate of investment pa = annual income from an investment of Rs.100 = $\dfrac{8}{110} \times 100 = \dfrac{80}{11} \simeq 7.27\%$

Exercises

1. How many shares of MV Rs.20/- each can be purchased for Rs.8080/- brokerage being 1%?

   (a) 400　　(b) 425　　(c) 440　　(d) 460　　(e) 470

2. Find the annual income got from Rs.5000/-, 10% stock at Rs.105/-.

   (a) Rs.420.3/-　　(c) Rs.476.2/-　　(e) Rs.570/-
   (b) Rs.460.8/-　　(d) Rs.500/-

3. A person invests in a 12% stock at Rs.120/-. What is the interest he obtains?

   (a) 10%　　(b) 11.5%　　(c) 12%　　(d) 12.67%　　(e) 13.25%

4. Find the annual income got by investing Rs.5950/- in 8% stock at Rs.119/-

   (a) Rs.400/-　(b) Rs.420/-　(c) Rs.430/-　(d) Rs.450/-　(e) Rs.470/-

5. Find the income derived from Rs.4480/- 12% stock at Rs.112/-

   (a) Rs.420/-　(b) Rs.460/-　(c) Rs.480/-　(d) Rs.500/-　(e) Rs.530/-

6. Rakesh invested Rs.5700/- in Rs.12/- shares quoted at Rs.9.50/- If the rate of dividend is 10%. What his annual income?

   (a) Rs.700/-　(b) Rs.720/-　(c) Rs.750/-　(d) Rs.770/-　(e) Rs.790/-

7. Kiran invested Rs.18,000/- in Rs.120/- shares of a company at 20% premium. If the company declares 6% dividend at the end of the year, how much does he get?

   (a) Rs.800/-　(b) Rs.820/-　(c) Rs.850/-　(d) Rs.900/-　(e) Rs.950/-

8. Find the income got from 90 shares of Rs.20/- each at 4 premium, brokerage being 1/2 per share are rate of dividend being 8% per annum. Find also the rate of interest on the investment.

   (a) 6.2%   (b) 6.4%   (c) 6.53%   (d) 6.9%   (e) 7.2%

9. Sekhar bought Rs.30/- shares in a company which pays 8% dividend. The money invested is such that it gives 9% on investment; At what price did he by the shares?

   (a) Rs.23.25/-   (c) Rs.25/-   (e) Rs.29/-
   (b) Rs.24.50/-   (d) Rs.26.67/-

10. To have an annual income of Rs.1500/- from a 10% stock at Rs.95/- the amount of stock needed is

    (a) Rs.14,200/-   (c) Rs.15,000/-   (e) Rs.17,000/-
    (b) Rs.14,850/-   (d) Rs.16,000/-

11. Govardhan invested some money in an 8% stock, at Rs.92/-. If Laxman wants to invest in an equally good 10% stock he must purchase a stock worth of

    (a) Rs.110/-   (b) Rs.115/-   (c) Rs.125/-   (d) Rs.130/-   (e) Rs.148/-

12. By investing in 15% stock at 75, one earns Rs.1800/-. What is the investment made?

    (a) Rs.8300/-   (c) Rs.8600/-   (e) Rs.9000/-
    (b) Rs.8400/-   (d) Rs.8800/-

13. A person invested Rs.1400/- in a stock at Rs.95/- to get an income of Rs.112/-. What is the dividend from the stock?

    (a) 7.5%   (b) 7.6%   (c) 8%   (d) 8.4%   (e) 9.2%

14. Bhatia invests same amount partly in 8% stock at Rs.104/- and partly in 13% stock at Rs.117/-. To obtain equal dividends from the investments what should be the ratio of the investments?

    (a) 11:10   (b) 12:11   (c) 13:9   (d) 14:9   (e) 15:7

Solutions

1. C.P. of each share = Rs.20/- + 1% of Rs.20/- = Rs.20.20

   No of shares = $\dfrac{8080}{20.20}$ = 400

2. Income from Rs.100/- stock = Rs.10/- MV = 105

   Income from Rs.5000/- stock = $\dfrac{10}{105}$ × 5000 = Rs.476.2

3. By investing Rs.120/- the income got = Rs.12/-

   By investing Rs.100/- the income = $\dfrac{12}{120}$ × 100 = Rs.10/-

   Interest 10%

4. By investing Rs.119/- income Rs.90/- = Rs.8/-

   By investing Rs.5950/- income = $\dfrac{8}{119}$ × 5950 = Rs.400/-

5. Income on Rs.100 stock = Rs.12/-

   Income on Rs.4480/- stock = $\dfrac{12}{112}$ × 4480 = Rs.480/-

6. No. of shares = $\dfrac{5700}{9.50}$ = 600

   Face value 600 × $Rs.12/-$ = Rs.7200/-

   Annual income = $\dfrac{10}{100}$ × 7200 = $Rs.720/-$

7. Cost of each stock = 120 + premium 20% of Rs.120/-

   = $120 + \dfrac{20}{100}$ × 120 = $Rs 144/-$

   No. of shares = $\dfrac{18000}{144}$ = $\dfrac{1000}{8}$ = Rs.125/-

   Face value = Rs.120/- × 125 shares = Rs.15,000/-

   Annual income = $\dfrac{6}{100}$ × 15000 = Rs.900/-

Stocks & Shares 235

8. Cost of share $= 20 + 4 + \dfrac{1}{2} =$ Rs. $\dfrac{49}{2}/-$

   Cost of 90 shares $= 90 \times \dfrac{49}{2} =$ Rs.2205/–

   Investment made = Rs.2205/-

   FV of 90 shares $= 90 \times 20 =$ Rs.1800/-; dividend on Rs.100/- is Rs.8/-

   $\therefore$ Dividend on Rs.1800/- $= \dfrac{8}{100} \times 1800 =$ Rs.144/–

   Income got = Rs.144/-

   Rate of interest on investment $= \dfrac{144}{2205} \times 100 = 6.53\%$

9. Let each share be purchased at Rs. x/-; dividend on Rs.30/- shares @ 8%

   $= \dfrac{8}{100} \times$ Rs.30/–

   Money on investment = Rs $x \times \dfrac{9}{100}$

   Since $\dfrac{8 \times 30}{100} = \dfrac{x \times 9}{100}$

   $\therefore x = \dfrac{8 \times 30}{9} = \dfrac{240}{9} = \dfrac{80}{3} = Rs.26.67/-$

10. For an income of Rs.10/- stock needed = Rs.100/-

    For an income of Rs.1500/- stock needed $= \dfrac{100}{10} \times 1500$

    $\qquad\qquad\qquad\qquad\qquad\qquad =$ Rs.15,000/-

11. For an income of Rs.8/- investment was Rs.92/-

    For an income of Rs.10/- investment should be $\dfrac{92}{8} \times 10 =$ Rs.115/-

12. To earn Rs.15/- investment = Rs.75/-

    To earn Rs1800/- investment $= \dfrac{75}{15} \times 1800$

    = Rs.9000/-

13. By investing Rs.1400/- income is Rs.112/-

By investing Rs.95/- income will be $\dfrac{112}{1400} \times 95$

$= \dfrac{38}{5} = 7.6\%$ (Remember: (Divided is always paid on FV) ; 7.6%)

14. For equal incomes of Rs.100/- say, let us find the investments. In the first option, for an income of Rs.100/- the investment $= \dfrac{104}{8} \times 100 = 1300$

In the second case, for an income of Rs.100/-

The investment $= \dfrac{117}{13} \times 100 = Rs.900/-$

Thus to get equal dividends the investments should be in the ratio of 1300: 900 i.e 13:9

Key

1)	a	2)	c	3)	a	4)	a	5)	c	6)	b
7)	d	8)	c	9)	d	10)	c	11)	b	12)	e
13)	b	14)	c								

32

True Discount

Definitions & Formulae

Let p_0 be the P.W, let the sum due be A after n y @ r pc pa. (see the example)

(P.W: Present Worth)

$$TD = \frac{p_0 nr}{100} \text{ (i) and } \therefore \text{ S.I on TD} = \left[\frac{\left(\frac{p_0 nr}{100}\right) n \times r}{100}\right] \text{ (ii)}$$

$\therefore$ S.I, on amount = TD + S.I on TD

$$= \frac{p_0 nr}{100} + \frac{p_0(nr)^2}{(100)^2} \text{ (iii)}$$

$\Rightarrow$ S.I on amount - TD = $\left[\frac{p_0(nr)^2}{100^2}\right]$

A sum of Rs. 1000/- at 10 pc pa will amount to $\left(1000 + \frac{1000 \times 3 \times 10}{100}\right)$ in 3 year i.e to 1000 + 300 i.e to 1300. So the payment of Rs. 1000/- now will clear a debt of Rs. 1300/- due in 3 years hence

Sum due (in 3 years) is 1300

Present Worth (P.W) is 1000

Def: True Discount (TD) = Amount due - P.W = 1300 - 1000 = 300

Note: Interest is reckoned on P.W and TD is reckoned on amount, we note the formulae:

1. P.W = $\frac{100 \times \text{Amount}}{(100 + rn)}$: $\left[1000 = \frac{100 \times 1300}{(100 + 10 \times 3)}\right]$

2. P.W = $\frac{100 \times TD}{rn}$: $\left[1000 = \frac{100 \times 300}{10 \times 3}\right]$

3. $TD = \dfrac{(P.W)\, r \times n}{100}$; $\left[300 = \dfrac{1000 \times 10 \times 3}{100}\right]$ equ(i)

4. $TD = \dfrac{Amount \times r \times n}{100 + (R \times n)}$; $\left[300 = \dfrac{1300 \times 10 \times 3}{100 + (10 \times 3)}\right]$

5. S.I (on Amount) $= \dfrac{1300 \times 10 \times 3}{100} = 390$

6. S.I on TD $= \dfrac{300 \times 10 \times 3}{100} = 90$

7. S.I (on Amount) - S.I on TD = 390 - 90 = 300 = TD

   i.e S.I (on Amount) - TD = S.I on TD   eqn (iii)

8. Amount $= \dfrac{S.I\,(on\,Amount) \times TD}{[(S.I\,on\,A) - TD)]}$ $\left[\dfrac{390 \times 300}{(390 - 300)} = \dfrac{390 \times 300}{90} = 13 \times\right.$

   $= 1300$

Let p_0 be the P.W. Let the sum be A after ny @ r pc pa

→ $TD = \dfrac{p_0 nr}{100}$ (i); → S.I on TD $= \left(\dfrac{p_0 nr}{100}\right) \dfrac{nr}{100}$

→ S.I on amount = TD + S.I on TD $= \dfrac{p_0 nr}{100} + \dfrac{p_1 (nr)^2}{(100)^2}$

→ $(S.I\,on\,Amount - TD) = p_0 \left(\dfrac{nr}{100}\right)^2$

→ Sum due (Amount) $= p_0 +$ Interest

$A = p_0 + p_0 \dfrac{rn}{100} = p_0 \left(1 + \dfrac{nr}{100}\right) = p_0 \dfrac{(100 + nr)}{100}$

Or $p_0 = \left(\dfrac{100A}{100 + rn}\right)$

$TD = \dfrac{p_0 rn}{100}$ ∴ $100\, TD = p_0 nr$

∴ $100\, TD + nr\, TD = p_0 nr + nr\, p_0 \dfrac{nr}{100}$

∴ $p_0 nr \left(1 + \dfrac{nr}{100}\right) = Anr$ ⇒ ∴ $A = \left[\dfrac{100TD}{nr} + TD\right]$

Exercises

1. What is the P.W of Rs. 2,100/- due in 4y hence @ 10 pc pa? Also find the true discount

   (a) 1300, 850   (c) 1400, 700   (e) 1500, 600
   (b) 1350, 800   (d) 1450, 650

2. The TD on a bill due in 4 years hence @ 10 pc pa is 600. Find the billed amount and its P.W

   (a) 2030, 1630   (c) 2050, 1600   (e) 2150, 1560
   (b) 2040, 1620   (d) 2100, 1500

3. What is the present worth of Rs. 14,000/- due 2y hence; rate being 20 pc pa?

   (a) 9200   (c) 9500   (e) 11,000
   (b) 9300   (d) 10,000

4. If the TD on an amount due 2y hence @ 20 pc pa is Rs. 4,000/-. What is the amount due, what is the P.W?

   (a) Rs. 12500/-, Rs. 9500/-   (d) Rs. 13900/-, Rs. 9800/-
   (b) Rs. 13000/-, Rs. 9600/-   (e) Rs. 14000/-, Rs. 10000/-
   (c) Rs. 13500/-, Rs. 9800/-

5. The TD on Rs. 1200/- due 2y hence is Rs. 200/-. What is r pc pa?

   (a) 9   (b) 10   (c) 11   (d) 12.5   (e) 14.75

6. The TD on a bill due 2y hence @ 10 pc pa is Rs. 200/-. Find the bill amount?

   (a) Rs. 1050/-   (c) Rs. 1175/-   (e) Rs. 1250/-
   (b) Rs. 1100/-   (d) Rs. 1200/-

7. The S.I on Rs. 1000/- for 2y is the same as the TD on Rs. 1200/- due 2y hence. The rate interest is r pc pa. What is r?

   (a) 10 pc pa   (c) 11.75 pc pa   (e) 13 pc pa
   (b) 10.5 pc pa   (d) 12.5 pc pa

8. If Rs. 200/- is the TD on a bill of Rs. 1200/- due at the end of a certain time, then what is the TD to be allowed on the same sum due at the end of double the time?

   (a) 328.56   (c) 342.86   (e) 344.50
   (b) 339.14   (d) 343.12

9. A shop keeper has two offers to sell a T.V offer A at a credit of Rs. 15,000/- to be paid after 2y; rate of interest being 10 pc pa (ii) offer B is Rs. 12,600/- cash. Which is the better offer? What is the profit bit there by?

   (a) A, Rs. 25/-   (c) A, Rs. 150/-   (e) A, Rs. 200/-
   (b) B, Rs. 98/-   (d) B, Rs. 100/-

10. X has to pay Y an amount Rs. 2300/- due 15 m hence, rate being 12 pc pa. Y has to pay X an amount Rs. 2600/- due 2y hence, rate being 10 pc pa. Who should give the larger amount?

    (a) X to Y Rs. 40/-   (d) Y to X; Rs. 120/-
    (b) X to Y Rs. 50/-   (e) Y to X Rs. 170/-
    (c) None to the other, Rs. 0/-

11. The P.W of Rs. 2781/- in 3 equal 4 monthly installments @ 9 pc pa. S.I is

    (a) Rs. 2265/-   (c) Rs. 2495/-   (e) Rs. 2625/-
    (b) Rs. 2385/-   (d) Rs. 2575/-

12. Sekhar purchases an electronic item for Rs. 9500/- he sells it for Rs. 12,000/- at a credit of 2 years. The rate is 10 pc pa. What is sekhar's gain or loss?

    (a) loss Rs. 390/-   (d) gain Rs. 500/-
    (b) loss Rs. 400/-   (e) gain Rs. 600/-
    (c) loss Rs. 50/-

13. A bill falls due in one year. It is agreed that one third amount is to be paid immediately and the remaining may be deferred for one year. With this arrangement the shopkeeper gains Rs. 100/-. What is the

bill amount, if the money is worth 10% (i.e rate of interest is 10 pc pa)?

(a) Rs. 3100/-   (c) Rs. 3300/-   (e) Rs. 3600/-
(b) Rs. 3200/-   (d) Rs. 3500/-

Solutions

1. $P.W = \dfrac{100 \times \text{Amount}}{100 + r \times n} = \dfrac{100 \times 2100}{100 + (10 \times 4)} = \dfrac{21 \times 10^4}{140} = \dfrac{3}{20} \times 10^4$
   $= 1500$

   TD = Amount due - (P.W) = 2100 - 1500 = 600

2. $TD = \dfrac{\text{Amount} \times r \times n}{(100 + r \times n)} \Rightarrow 600 = \dfrac{A \times 10 \times 4}{(100 + 10 \times 4)} = \dfrac{40A}{140}$

   $\therefore A = \dfrac{600 \times 140}{40} = $ Rs. 2100/-; from def of TD.

   P.W = (Amount due) - TD = 2100 - 600 = 1500

3. Amount = 14000; r = 20 pc pa; n = 2y;

   $P.W = \dfrac{100 \times \text{Amount}}{(100 + r \times n)} = \dfrac{100 \times 14 \times 10^3}{(100 + 20 \times 2)} = \dfrac{14 \times 10^5}{140} = $ Rs. 10^4

4. Amount = $TD \dfrac{[100 + r \times n]}{(r \times n)} = 4000 \dfrac{[100 + 20 \times 2]}{20 \times 2} = 14,000$

   $P.W = \dfrac{TD \times 100}{(r \times n)} = \dfrac{4000 \times 100}{20 \times 2} = 10,000$

5. A = Rs. 1200/-; TD = Rs. 200/-; P.W = p_0 = 1200 - 200 = 1000 and
   $TD = 200 = \dfrac{1000 \times r \times 2}{100}$ from (3)

   $\therefore r = 10$ pc pa

6. TD = Rs. 200/- ; n = 2y; r = 10 pc pa; A = ?

$$\therefore 200 \text{ (TD)} = \frac{A \times 10 \times 2}{(100 + 2 \times 10)} \text{ and } \therefore A = \frac{200 \times 120}{20} = 1200$$

7. S.I on Rs. 1000/-; (P.W) = $\frac{1000 \times 2 \times r}{100}$ = TD on Rs. 1200/- (A)

$$= \frac{1200 \times 2r}{(100 + 2r)} \quad \therefore 10 = \frac{1200}{(100 + 2r)}; \ 20r = 200$$

$\therefore r = 10$ pc pa

8. TD = 200; sum (A) = Rs. 1200/-; r = ? n = ?

$$200 = \frac{1200 rn}{(100 + rn)} \quad \therefore (100 + rn) = 6rn \Rightarrow rn = 20;$$

$$TD_2 = \frac{1200 r.2n}{(100 + r.2n)} = \frac{2400 \times 20}{100 + 2 \times 20} = 342.86$$

9. Offer A: A = 15,000; due in n = 2y, r = 10 pc pa

$$(P.W) = \frac{100 \times \text{Amount}}{(100 + nr)} = \frac{100 \times 15000}{(100 + 2 \times 10)} = 12,500$$

Offer B = 12,600 cash: offer B is better

Profit = offer B - offer A = 12,600 - 12,500 = Rs. 100/-

10. X to pay Y :

$$P.W = \frac{100A}{(100 + r \times n)} = \frac{100 \times 2300}{(100 + \frac{5}{4} \times 12)} = \frac{23 \times 10^4}{115} = 2000$$

Y to pay X :

$$(P.W) = \frac{100A}{(100 + r \times n)} = \frac{100 \times 2600}{100 + 2 \times 10} = 2170$$

So y has to give (2170 - 2000) = Rs. 170/- more to X

True Discount 243

11. Each installment is $\frac{2781}{3}$ = Rs. 927/-

(a) (P.W) of first installment is Rs. 927/- due in 4m (or $\frac{1}{3}y$) hence @ 9 pc pa rate = $\frac{100 \times 927}{(100 + \frac{1}{3} \times 9)} = \frac{92700}{103} = $ Rs. 900/-

(b) P.W 2nd installment of Rs. 927/- due in 8m (or $\frac{2}{3}y$) with r = 9 pc pa interest is $\left[\frac{100 \times 927}{(100 + \frac{2}{3} \times 9)}\right] = $ Rs. 874.53/-

(c) P.W of 3rd installment of Rs. 927/- due in one year with r = 9 pc is $\frac{100 \times 927}{(100 + 1 \times 9)} = \frac{100 \times 927}{109} = $ Rs. 850.46/-

So, the (P.W) of Rs. 2781/- is $\begin{cases} 900 \text{ from (a)} \\ + 874.53 \text{ (b)} \\ + 850.46 \text{ (c)} \\ \overline{2624.99} \end{cases}$; Rs. 2625/-

12. CP = 9500

P.W of Rs. 12000/- due in 2y;

r = 10 pc pa; is = $\frac{12000 \times 100}{(100 + 10 \times 2)} = $ Rs. 10000/-

Thus S.P = 10,000/-; gain = 10,000 - 9500 = Rs. 500/- gain

13. Let Rs. x/- be the amount due in one year;

r being 10pc pa

$\frac{x \times 100}{(100 + 10 \times 1)} = \frac{10x}{11}$

Actual payment agreement (i) $\frac{x}{3}$ immediate payment (ii) $2\frac{x}{3}$ due in one year, r = 10 pc pa

Its (P.W) = $\dfrac{2x}{3} \times \dfrac{100}{(100 + 10 \times 1)}$

Total actual payment = $\dfrac{x}{3} + \dfrac{20x}{33} = \dfrac{31x}{33}$

Against $\dfrac{10x}{11}$ or $\dfrac{30x}{33}$

Gain = $\dfrac{x}{33}$ = 100 (given)

∴ x = Rs. 3300/-

Key

1)	e	2)	d	3)	d	4)	e	5)	b	6)	d	7)	a
8)	c	9)	d	10)	e	11)	e	12)	d	13)	c		

33

Banker's Discount

Definitions and formulae

If an organization x purchases goods worth say Rs.50,000/- from another y at a credit of say 1year, y gives a bill of exchange' to x. X signs it and authorizes y to with draw the money from its bank account after 1year. This date (after one year) is called the 'nominal due' date. A grace period of three days is added to get a 'legally due' date

If y draws the money before the legal due date he can get the money from the banker which deducts to S.I on the 'face value' (Rs.50,000/- in this example) for the period between the discounted date of the bill and the legal due date. This amount is known as '**Banker discount**' (defined as the S.I for the period between the discounted date and legal due date).

Banker's gain = BD - TD for the unexpired time

1. BD = S.I on the bill for the unexpired time = $\dfrac{\text{Amount} \times \text{rate} \times \text{time}}{100}$

2. BG = BD - TD = S.I on TD (as shown below)

   BG = BD - TD

   $= \dfrac{Anr}{100} - \left(\dfrac{Anr}{100+nr}\right) = \dfrac{100Anr + An^2r^2 - 100Anr}{100(100+nr)}$

   $= \left(\dfrac{Anr}{(100+nr)}\right) \dfrac{nr}{100}$ i.e TD. $\dfrac{nr}{100}$ = S.I on TD

   Thus (2) BG = BD - TD = S.I on TD = $\dfrac{An^2r^2}{100(100+nr)}$ (2)

   Now P.W = $\dfrac{100A}{100+nr}$

$$\therefore \frac{TD^2}{P.W} = \frac{A^2n^2r^2}{(100+nr)^2} \cdot \frac{1}{P.W} = \frac{A^2n^2r^2}{(100+nr)^2} \cdot \frac{(100+nr)}{100A}$$

$$\frac{An^2r^2}{(100+nr)100} = BG \text{ from (2)}$$

$$\therefore \text{ Thus (3) } BG = \frac{TD^2}{P.W} = \text{ or } TD = \sqrt{(P.W)(BG)} \quad (3)$$

Now $BD \times TD = \dfrac{Anr}{100} \cdot \dfrac{Anr}{100+nr}$; also $BD - TD = BG = \dfrac{An^2r^2}{(100+nr)100}$ from (2)

$$\frac{BD \times TD}{(BD-TD)} = \frac{Anr}{100} \cdot \frac{An}{(100+n)} \cdot \frac{(100+n)100}{An^2r^2} = A$$

$$\rightarrow \text{ Thus } A = \frac{BD \times TD}{BD-TD} = \frac{BD \times TD}{BG} \quad (4)$$

Now $TD = \dfrac{TD^2}{TD}$; Also $P.W = \dfrac{100(TD)}{nr}$

From this $TD = (P.W) \dfrac{nr}{100}$

Hence $TD = \dfrac{TD^2}{TD} = \dfrac{TD^2}{(P.W)(nr)} \times 100$. Also we have seen above

$$\frac{TD^2}{(P.W)} = BG \text{ (3) above}$$

$$\therefore TD = \frac{BG \cdot 100}{nr} \quad (5)$$

Exercises

1. The TD on a bill of Rs.6000 is Rs.1000/- what is the BD? What is BG?

   (a) Rs.1200; 200/- (c) Rs.1300, 250/- (e) Rs.1390, 350/-
   (b) Rs.1250, 220/- (d) Rs.1340, 330/-

2. The TD on a certain sum due 2y @ 10 pc pa rate is Rs.1000/- What is the BD on the sum for the same time, at the same rate?

   (a) Rs.1050/-
   (b) Rs.1120/-
   (c) Rs.1200/-
   (d) Rs.1260/-
   (e) Rs.1300/-

3. The (P.W) of a bill due some time hence is Rs.5000/- TD on the bill is Rs.1000/- what is the BD; and BG?

   (a) Rs.985/- Rs.140/-
   (b) Rs.1000/-, Rs.150/-
   (c) Rs.1100/-, Rs.150/-
   (d) Rs.1200/-, Rs.200/-
   (e) Rs.1300/-, Rs.300/-

4. The (P.W) of a certain bill due some time hence is Rs.5000/- and the TD is Rs.1000/- What is the BD?

   (a) Rs.1100/-
   (b) Rs.1120/-
   (c) Rs.1150/-
   (d) Rs.1180/-
   (e) Rs.1200/-

5. The BG on a certain sum due 2y hence is 1/6 BD. What is r pc pa?

   (a) 9.5 pc pa
   (b) 10 pc pa
   (c) 10.75 pc pa
   (d) 11.25 pc pa
   (e) 12.5 pc pa

6. The BD on a sum of money for 2y is Rs.6000/- The TD on the same sum for 2y is Rs.5000/- What is r pc pa?

   (a) 9 pc pa
   (b) 9.75 pc pa
   (c) 10 pc pa
   (d) 11.5 pc pa
   (e) 12.5 pc pa

7. The BD on Rs.5000/- @ 10 pc pa is equal to the TD on Rs.6000/- for the same time at the same rate, what is the time?

   (a) 1 year
   (b) $1\frac{1}{4}$ year
   (c) $1\frac{1}{2}$ year
   (d) 2 years
   (e) 5 years

Quantitative Aptitude

8. The BD on a bill due $\frac{1}{2}$ year hence @ 12 pc pa is Rs.1272/- What is the TD?

   (a) Rs.1075/-   (b) Rs.1090/-   (c) Rs.1120/-
   (d) Rs.1175   (e) Rs.1200/-

9. The BD on a sum Rs.6000/- is Rs.1200/- what is the TD? What is the BG?

   (a) Rs.920/-, Rs.275/-   (d) Rs.980/-, Rs.250/-
   (b) Rs.940/-, Rs.280/-   (e) Rs.1000/-, Rs.200/-
   (c) Rs.960/-, Rs.300/-

10. The BD on Rs.8000/- @ 10 pc pa is the same as the TD on Rs.9600/- for the same time at the same rate. The time is

    (a) 1.75y   (b) 2y   (c) 2.25y   (d) 3y   (e) 3.5y

Solutions

1. $TD = \dfrac{Arn}{(100 + rn)} = \dfrac{6000 \times rn}{(100 + rn)} = 1000$ (given)

   $\therefore 10^5 + 1000rn = 6000rn \quad \therefore 5000rn = 10^5 \quad \therefore rn = 20$

   Now $BD = \dfrac{Arn}{100} = \dfrac{6000 \times 20}{100} = 1200$

   $BG = BD - TD = 1200 - 1000 = 200$

2. $TD = 1000 = \dfrac{A \times 10 \times 2}{(100 + 10 \times 2)} \quad \therefore A = \dfrac{120 \times 1000}{20} = 6000$

   $BD = \dfrac{Arn}{100} = \dfrac{6000 \times 10 \times 2}{100} = 1200$

3. TD = Rs.1000/- (P.W) = Rs.5000/-;

   $BG = \dfrac{TD^2}{(P.W)} = \dfrac{10^6}{5000} = 200;$

   $BD = TD + BG = 1000 + 200 = 1200$

4. (P.W) = Rs.5000/-; TD = Rs.1000/-;

   $BG = \dfrac{TD^2}{(P.W)} = \dfrac{10^6}{5 \times 10^3} = 200;$

   $BD = BG + TD = 200 + 1000 = 1200$

5. $BG = \dfrac{1}{6} BD;\ TD = BD - BG = BD - \dfrac{1}{6} BD = \dfrac{5}{6} BD$

   We know $TD = \dfrac{BG.100}{rn};\ n = 2y;$

   $\therefore \dfrac{5}{6} BD = \dfrac{1}{6} BD . \dfrac{100}{rn};$

   $\therefore r = \dfrac{100}{2 \times 5} = 10$ pc pa

6. $BD = 6000 = \dfrac{Ar.2}{100} \quad \therefore Ar = 3 \times 10^5;$

   $TD = \dfrac{Ar \times 2}{(100 + 2r)} = 5000 \quad \therefore (100 + 2r) = \dfrac{Ar \times 2}{5000}$

   $= \dfrac{(3 \times 10^5) \times 2}{5000} = \dfrac{600}{5} = 120$

   $\therefore 2r = 120 - 100 = 20 \quad \therefore r = 10$ pc pa

7. $BD = \dfrac{5000 \times 10n}{100} = 500n;\ TD = 500n = \dfrac{6000 \times 10n}{(100 + 10n)}$

   $\Rightarrow (100 + 10n) = \dfrac{60000n}{500n} = 120 \quad \therefore n = 2y$

8. $BD = Rs.1272/- = A \times \dfrac{1}{2} \times \dfrac{12}{100}$ $\therefore A = 1272 \times \dfrac{200}{12} = 21,200$

$$TD = \left[\dfrac{21200 \times \frac{1}{2} \times 12}{\left(100 + \frac{1}{2} \times 12\right)}\right] = \dfrac{21200 \times 6}{106} = Rs.1200/-$$

9. $BD = 1200 = \dfrac{6000\,nr}{100}$

$\therefore nr = \dfrac{12 \times 10^4}{6 \times 10^3} = 20$; from 4, p34.1

$TD = \left[\dfrac{6000 \times nr}{(100 + nr)}\right] = \dfrac{6000 \times 20}{(100 + 20)} = 1000$

$BG = BD - TD = 1200 - 1000 = 200$

10. $BD = \dfrac{8000 \times n \times 10}{100} = 800n;\ TD = \dfrac{9600 \times n \times 10}{(100 + 10n)}$

$= \dfrac{96 \times 10^3 n}{(100 + 10n)}$; Given $800.n = \dfrac{96 \times 10^3 n}{[100 + 10n]}$

$\therefore (100 + 10n) = \dfrac{96 \times 10^3}{800} = 120$

$\therefore 10n = 20$ and $n = 2y$

Key

1)	a	2)	c	3)	d	4)	e	5)	b
6)	c	7)	d	8)	e	9)	e	10)	b

34

Non Conforming Items

Exercises

Find the non conforming item (s) in the following groups (The groups are classified in to easily identifiable groups following certain trends)

Divisibility by simple numbers:

1. 6, 9, 15, 21, 24, 28, 30, 33....

2. 35190, 35196, 35202, 35208, 35214, 35216, 35226, 35232, 35238....

3. 129, 132, 143, 154, 165, 176, 187, 189, 220....

4. 371718, 2721797, 372696, 3721685, 371674, 3721663, 3721653, 3721641, 3721630....

5. 41652, 41640, 41628, 41616, 41604, 41592, 41588....

Arithmetic Progression & Geometric Progression

6. 71, 75, 70, 83, 87, 91, 95, 99, 102, 107, 111, 115, 119....

7. $1, \dfrac{1}{4}, \dfrac{1}{7}, \dfrac{1}{10}, \dfrac{1}{13}, \dfrac{1}{16}, \dfrac{1}{18}, \dfrac{1}{22}, \dfrac{1}{25}, \dfrac{1}{28}$ ....

8. $\dfrac{1}{13}, \dfrac{1}{26}, \dfrac{1}{39}, \dfrac{1}{52}, \dfrac{1}{64}, \dfrac{1}{78}, \dfrac{1}{91}, \dfrac{1}{104}, \dfrac{1}{117}$ ....

Higher order functions like $(an^2 \pm 1)$

9. 901, 962, 1024, 1090, 1157....

10. 899, 960, 1023, 1088, 1155, 1226, 1295, 1368....

11. 243, 289, 339, 393, 451, 513, 578....

(B) Insert the missing number: (i) Divisibility problems

12. 5082 1632, 5082 1648, 5082 1664, 5082 1680, ..., 5082 1712....

13. 695600, 695680, 695760, 695840, ...

Progressions

14. 27, 81, 243, 729, 2187, ...19683, 59049, 177147....

15. 17, 119, 833, 5831, 40817, ..., 2000033,....

16. $\dfrac{1}{17}, \dfrac{1}{29}, \dfrac{1}{41}, \dfrac{1}{53}, \dfrac{1}{65}, \dfrac{1}{77}, \ldots \ldots \dfrac{1}{101}, \dfrac{1}{113}, \ldots$

17. $\dfrac{1}{23}, \dfrac{1}{42}, \dfrac{1}{61}, \dfrac{1}{80}, \dfrac{1}{99}, \dfrac{1}{137}, \dfrac{1}{156}, \ldots$

Functions like $an^2 \pm 1$

18. $\dfrac{1}{31}, \dfrac{1}{38}, \dfrac{1}{45}, \dfrac{1}{52}, \dfrac{1}{59}, \dfrac{1}{66}, \dfrac{1}{73}, \dfrac{1}{80} \ldots \ldots \dfrac{1}{94}, \dfrac{1}{101}, \dfrac{1}{108}$

19. 290, 325, 362, 401, 442, 485....

20. 42874, 46655, 50652, ..., 59318, 63999....

($m^k \times n$) type

21. 256, 512, 1024, 2048,....8192

22. 414, 828, 1650, 3312, 6624, 13248, 26496....

23. 1226, 2452, 4904, 9808, 19616....

(m^k n+p) type

24. 258, 514, 1026, 2050....8194....

25. 416, 830, 1658, 3314, 6626, 13250....

Solutions

1. 28 since (28) is not divisible by 3

2. (35216) not divisible by 6 (ie by 2 and by 3)

3. (189) not divisible by 11. other are 1 + 9 = 10; 8 difference not divisible by 11; or not equal to zero

4. (3721653) not divisible by 12. Other are sum of digits = 27 divisible by 3; but 53 is not divisible by 4.

5. (41588) not divisible by 3. Since sum of digits = 26; but divisible by 4, so not divisible by 12.

6. (102) is the 'odd' term since the given numbers chain is AP; First term 'a' = 71, common differenced = 4

7. $\left(\dfrac{1}{18}\right)$ is non conforming denominators are in AP 'a' = 1; 'd' = 3

8. $\left(\dfrac{1}{64}\right)$ is non conforming other denominators are multiples of 13.

9. (1024) is non conforming other terms are of the form $an^r \pm 1$; a = 1; n = 30; 31, 32, 33, 34.... an AP (with 1st term 30; CD = 1;) r = 2; and (+1): Thus, the term are $1.30^2 + 1$; $1.31^2 + 1$; $\boxed{1.32^2 + 1}$; $1.33^2 + 1$ $34^2 + 1$....

10. (1226) is not conforming. 6 the terms are $a.n^2 - 1$; a = 1; n changes from 3: 30, 31, 32, 33, 34, 35, 36, 37....

11. (578) is not conforming. Other number are $2n^2 + 1$ with n = 11, 12, 13, 14, 15, 16 and ⑰

The missing numbers are:

12. 50821696 Divisibility by 16

13. 695920 Divisibility by 80

14. (6561) GP 1st term 27, CR 3

15. (285719) GP 1st term 17, CR 7

16. $\left(\dfrac{1}{89}\right)$ Denominators are in AP 1st term 17, CD 12

17. $\left(\dfrac{1}{118}\right)$ Denominators are in AP 1st term 23, CD 19

18. $\left(\dfrac{1}{87}\right)$ Denominators are in AP 1st term 31, CR 7

19. (530) no's (n^2+1) with n = 17, 18, 19, 20, 21, 22, (23)

20. (63999) no's $(2n^2+1)$ with n = 35, 36, 37, 38, 39, (40)

21. (4096) no's $m^k n$ with m = 2, k = 1, 2, 3, 4, (5), 6 and n = 128

22. (52992) no's $m^k n$; m = 2, k = 1, 2, 3, 4, 5, 6, 7, 8, ...n = 207

23. (39232) no's $m^k n$; m = 2, k = 1, 2, 3, 4, 5, 6, ...; n = 613

24. (4098) no's $m^k n$ + 2; m = 2; k = 1, 2, 3, 4, 5, 6; n = 128

25. (26498) no's $m^k n$ + 2; m = 2; k = 1, 2, 3, 4, 5, 6

35

Permutations & Combinations

Introduction

Number of 'permutations' (different ordered arrangements of a given number of things by taking some or all at a time) of n things, taken r at a time is given by $n_{p_r} = \dfrac{\lfloor n}{\lfloor n-r} = n(n-1)(n-2)........(n-r+1)$

No of permutations of n objects (of which a_1 are alike (of one kind) a_2 are alike (of another kind) etc & a_r are alike of r th kind) is

$$\dfrac{\lfloor n}{\lfloor a_1 \; \lfloor a_2 \; .... \; \lfloor a_r}$$

A **combination** is a un ordered group or selection which can be formed by taking some or all of a number of objects

The **permutations** made with letters x, y, z taking all at a time are six. x y z, x z y, y z x, y x z, z x y and z y x. The **only combination** that can be formed of letters x, y, z taking all at a time is x y z.

X y & y x are two different **permutations** but they represent the same **combination**

The no of all combinations of n things taken r at a time is $n_{c_r} = \dfrac{\lfloor n}{\lfloor r \; \lfloor n-r}$

→ $n_{c_r} = n_{c_{(n-r)}}$

Note: $\lfloor 0 = 1$

Exercises

1. In how many ways can the letters of the word ('NEEDLE') be arranged?

   (a) 30　　(b) 60　　(c) 120　　(d) 720　　(e) 1480

2. How many words with or without meaning can be formed using the letters of the word NAGPUR using each letter only once?

   (a) 360 (b) 720 (c) 1440 (d) 4800

3. In how many ways the letters of the word 'ARRANGE' be arranged so that the vowels always come together?

   (a) 60 (b) 360 (c) 720 (d) 1440

4. In how many different ways the letters of EDUCATION be arranged in such a way that the vowels always come together?

   (a) 120 (b) 480 (c) 1440 (d) 14400

5. How many words can be formed from the letters of the word PORCELAIN so that the vowels are never together?

   (a) 360 (b) 480 (c) 720 (d) 720 × 480

6. E T O R V A is the name of a medical Tablet. In how many ways can the letters in this word be arranged so that the vowels occupy only the odd positions?

   (a) 6 (b) 24 (c) 36 (d) 48

7. In how many different ways can the letters of the word EMERGE be arranged in such a way that the vowels occupy only the odd positions?

   (a) 1 (b) 3 (c) 6 (d) 36

8. In how many ways the letters of the word 'C A M B R I D G E' be arranged so that the vowels occupy only the odd positions?

   (a) 60 (b) 360 (c) 2160 (d) 21600

9. From a group of 8 boys & 8 girls, five are to be selected to form a music team, so that at least 2 girls are there in the team. In how many ways this can be done?

   (a) 3000 (b) 3136 (c) 3144 (d) 3200

10. An urn contains 3 white balls, 3 red balls & 5 green balls. In how many ways can four balls be drawn from the urn, so that at least one green ball is to be included in the drawn?

    (a) 65   (b) 100   (c) 150   (d) 315

11. How many 4 digit numbers can be formed using the digits 6, 8, 4, 1, 7, 3, 0? No digit is to be repeated. The 4 digit no should end with zero.

    (a) 24   (b) 60   (c) 120   (d) 720

12. A box contains 3 white balls, 3 black balls, 3 red balls. In how many ways can 3 balls be drawn from the box such that at least one black ball is to be included in the draw?

    (a) 52   (b) 64   (c) 76   (d) 82

13. How many words can be formed from the letters of the word SIMPLE so that the vowels are never together?

    (a) 120   (b) 240   (c) 480   (d) 720

14. How many four digit numbers can be formed from the 7 digits 1, 3, 7, 9, 0, 2, 4 which are divisible by 5 & none of the digits is repeated?

    (a) 60   (b) 120   (c) 240   (d) 4800

15. In how many different ways can the letters of the word SANCTION be arranged, in such a way that the vowels always come together?

    (a) 540   (b) 1080   (c) 2160   (d) 3240

16. In how many different ways can the letters of the word RANKING be arranged so that the vowels always come together?

    (a) 180   (b) 360   (c) 720   (d) 1440

17. In how many different ways can the letters of the word OPERATION be arranged so that the vowels always come together?

    (a) 1800   (b) 3600   (c) 7200   (d) 14,400

Solutions

1. The word contains 1n, 3E, 1D, 1L: 6letters (one repeated thrice)

    Required number of ways $= \dfrac{\lfloor 6}{\lfloor 1 \; \lfloor 3 \; \lfloor 1 \; \lfloor 1} = \dfrac{6.5.4.3.2.1}{3.2.1} = 120$

    [Elaborating: □ □ □ □ □ □ The six letters N, E, E, D, L, E can be arranged in $6 \times 5 \times 4 \times 3 \times 2 \times 1 \; \lfloor 6$ ways.

    The repeated letter E can be arranged in □ □ □ $3 \times 2 \times 1 \; \lfloor 3$ ways.

    Hence the required no of ways $= \dfrac{\lfloor 6}{\lfloor 3}^{3\;2\;1} = 6.5.4 = 120$

    (In general the **rule** is: If there are 'n' objects, of which 'a', are alike (of one kind), a_2 are alike (second variety)....& a_r are all same, of rth variety; such that $a_1 + a_2 + \ldots + a_r = n$, the number of permutations of these n objects is $\dfrac{\lfloor n}{\lfloor a_1 \; \lfloor a_2 \; \ldots \; \lfloor a_k}$

2. The word contains 6 letters NAGPUR □ □ □ □ □ □

    The first space can be filled in 6ways, second in 5ways & so on

    ∴ The Reqd no of words $= \lfloor 6 = 6 \times 5 \times 4 \times 3 \times 2 \times 1 = 720$ ways

3. "ARRANGE" contains 7 letters.

    The vowels are A, A & E (Three).

    The consonant R is repeated twice.

    Considering the 3 vowels (AAE) together as one entity the letters are

    R, R (Repeated twice) N, G, $\left(\text{AAE} \right)$
    1 2   3 4   5

    No of ways of arranging there 5 "letters" $= \dfrac{\lfloor 5}{\lfloor 2} = \dfrac{5.4.3.2.1}{1.2} = 60$

    The three vowels AAE can be arranged among themselves in $\lfloor 3 = 3.2.1 = 6$ways

    Required number of ways $= 60 \times 6 = 360$

4. There are 9 letters in the word, EUAIO five vowels & four consonants D C T N. Treating (E U A I O) as one entity the five letters.

D C T N (E U A I O) can be arranged in $\lfloor 5$ ways = 5.4.3.2.1

1 2 3 4   5   = 120 ways

The vowels (E U A I O) can be arranged among themselves in $\lfloor 5$ = 5.4.3.2.1 = 120 ways

Required number of ways = 120 × 120 = 14400 ways

5. The word contains 9 letters; consonants are P R C L N (five) vowels are O E A I (four).

Taking the four vowels (O E A I) together as one entity the six letters P R C L N (O E A I)

1 2 3 4 5   6

Can be arranged in $\lfloor 6$ ways = 6.5.4.3.2.1 = 720 ways

The four vowels may be arranged among themselves in $\lfloor 4$ = 4.3.2.1 = 24 ways

Thus the number of words each having the **vowels together** = 720 × 24 ways - (1)

The total number of words that can be formed by using all the nine letters of the word PORCELAIN = $\lfloor 9$ ways

= 9.8.7.6.5.4.3.2.1 = 504 × 720 ways

Hence the number of words each having the **vowels never together** is from (i) & (ii)

= [(504) × (720) - (720 × 24) ways]

= 720 (504 - 24) = 720 (480) ways

6. The word contains 3 vowels (In the alphabetical (Dictionary) order. A E O & 3 Consonants R T V

☐ ☐ ☐ ☐ ☐ ☐
1 2 3 4 5 6

The three vowels can be placed in any of threes lots 1, 3, 5

The number of ways of arranging the vowels is

$$3p_3 = \frac{\lfloor 3}{\lfloor 3-3} = \frac{3.2.1}{\lfloor 0} = \frac{6}{1} = 6 \text{ ways}$$

The 3 consonants can be arranged in slots 2, 4, 6 again in

$$3p_3 = \frac{\lfloor 3}{\lfloor 3-3} = \frac{3.2.1}{1} = 6 \text{ ways}$$

Thus the required number of ways = $6 \times 6 = 36$ ways

If proof is needed. **Check:-**

Filling vowels

Slots	1	3	5	
	A	E	O	V_1
	A	O	E	V_2
	E	O	A	V_3
	E	A	O	V_4
	O	A	E	V_5
	O	E	A	V_6

There are 6 ways of filling vowels in odd slots. Named V_1 through V_6

In each of there arrangements there are six ways of filling even slots with consonants.

Arrangement V1:

	A	☐	E	☐	O	☐
		2		4		6
C_1		R		T		V
C_2		R		V		T
C_3		T		V		R
C_4		T		R		V
C_5		V		R		T
C_6		V		T		R

Thus for vowel arrangement V_1 there are six consonant arrangements $V_1C_1, V_1C_2, V_1C_3, V_1C_4, V_1C_5$ & V_1C_6

Similarly for each other vowel arrangement $V_2, V_3....V_6$ there will be 6 consonant arrangements

For eg: for V_5: we have

$V_5C_1, V_5C_2, V_5C_3, V_5C_4, V_5C_5, V_5C_6$

Total = $6 \times 6 = 36$ ways

7. The word has 3 vowels E, E, E & 3 consonants M R G

    Let the slots be marked us 1 to 6

    ☐ ☐ ☐ ☐ ☐ ☐
    1   2   3   4   5   6

    The three (repeated) vowels E, E, E, should be place in slots 1, 3, 5

    Consonants in the slots 2, 4, 6

    The (3) consonants can fill the (3) even slots in $3p_3$ ways

    $= \dfrac{\lfloor 3}{\lfloor 3-3} = \dfrac{\lfloor 3}{\lfloor 0} = \lfloor 3 = 3.2.1 = 6$ way

    The 3 repeating vowels (E'S) can fill the 3 odd positions in $\dfrac{3p_3}{3p_3} = \dfrac{6}{6}$

    = 1 way

    ∴ The Required number of ways = $6 \times 1 = 6$

8. There are total nine letters

    There are 3 vowels A E I (In the alphabetical order)

    6 consonant B C D G M R

    There are five odd numbered

    Slots & four even numbered slots

    The three vowels can be placed at any of the three places, out of the five slots marked 1, 3, 5, 7 & 9

    The number of ways of arranging the vowels = $5p_3$

    $= \dfrac{\lfloor 5}{\lfloor 5-3} = \dfrac{5.4.3.2.1}{2.1}$ 60 ways (i)

    The 6 consonants can be placed at any of the four even numbered slots 2, 4, 6, 8.

    The number of ways for this is $6p_4 = \dfrac{\lfloor 6}{\lfloor 6-4} = \dfrac{6.5.4.3.2.1}{2.1}$ 360 ways

    Hence the required number of ways = $60 \times 360 = 21600$

Combinations

9. The team may consist a) 3 boys & 2girls or (b) 2 boys & 3 girls (c) or 1 boy & 4 girls or (d) all the five girls.

The required number of ways is the summation of

a) the first combination: $8c_3 \times 8c_2$ b) the second: $8c_2 \times 8c_3$
c) the third: $8c_1 \times 8c_4$ & d) the fourth: $8c_0 \times 8c_5$

The required number is (a, b) $\dfrac{2 \times |\underline{8}}{|\underline{3}\ |\underline{5}} \times \dfrac{|\underline{8}}{|\underline{2}\ |\underline{6}} + (c)$

$\dfrac{|\underline{8}}{|\underline{1}\ |\underline{7}} \times \dfrac{|\underline{8}}{|\underline{4}\ |\underline{4}} + (d)\, 1 \cdot \dfrac{|\underline{8}}{|\underline{5}\ |\underline{3}}$

$= (a, b)\, 2 \times \dfrac{8.7.6.5.4.3.2.1}{3.2.1.5.4.3.2.1} \times \dfrac{8.7.(6.5.4.3.2.1)}{1.2.6.5.4.3.2.1} + (c)\, 8 +$

$(d)\, \dfrac{8.7.6}{3.2.1} = \dfrac{336}{3} \times 28 + 8 + 56$

$= 112 \times 28 + 8 + 56$

$= 3136 + 8 + 56$

$= 3200$

10. We may draw

 (a) 1 green ball from 5 green balls & three non green from 3 + 3 i.e 6 non green balls
 (b) 2 green balls from 5g & 2 non green from 6ng
 (c) 3 green balls from 5g & 1 ng from 6 ng balls
 (d) 4 green balls from 5g

Required result is a + b + c + d

(a) $5c_1 \times 6c_3$ (b) $5c_2 \times 6c_2$ (c) $5c_3 \times 6c_1$ (d) $5c_4$

(a) $\dfrac{|\underline{5}}{|\underline{1}\ |\underline{4}} \times \dfrac{|\underline{6}}{|\underline{3}\ |\underline{3}} = 5 \times \dfrac{6.5.4}{1.2.3} = 5 \times 20 = 100$

(b) $\dfrac{|\underline{5}}{|\underline{2}\ |\underline{3}} \times \dfrac{|\underline{6}}{|\underline{2}\ |\underline{4}} = \dfrac{5.4}{1.2} \times \dfrac{6.5}{1.2} = 10 \times 15 = 150$

(c) $\dfrac{\lfloor 5}{\lfloor 3 \lfloor 2} \times \dfrac{\lfloor 6}{\lfloor 1 \lfloor 5} = \dfrac{5.4}{1.2} \times 6 = 60$

(d) $\dfrac{\lfloor 5}{\lfloor 4 \lfloor 1} = 5$

Required number is $100 + 150 + 60 + 5 = 315$

11. Total number of digits given is 7

    We must have the numbers 0 is ☐ ☐ ☐ ☐
                                               4   3   2   1

    Slot 1 right most, In slot 2 any of the remaining six digits

    This is done in 6 ways. slot 3 by the remaining five digits.

    In 5 ways; similarly slot 4 in four ways

    Thus the required no of 4 digit numbers are $1 \times 6 \times 5 \times 4 = 120$ (c)

12. Total 9 balls; 3 black balls. The required outcome is the sum of

    (i) One black ball out of three black ball & two out of the remaining six non black balls

    (ii) Two black ball out of three black ball 8 one out of the other six

    (iii) All three black ball from 3bb (plus no other coloured ball)

    The number of ways for

    (i) $3c_1 \times 6c_2$ for

    (ii) $3c_2 \times 6c_1$

    (iii) $3c_3 \times 6c_0$

    $= \dfrac{\lfloor 3}{\lfloor 1 \lfloor 2} \times \dfrac{\lfloor 6}{\lfloor 2 \lfloor 4} \times \dfrac{\lfloor 3}{\lfloor 2 \lfloor 1} \times \dfrac{\lfloor 6}{\lfloor 1 \lfloor 5} + \dfrac{\lfloor 3}{\lfloor 3 \lfloor 0} \dfrac{\lfloor 6}{\lfloor 0 \lfloor 6}$

    $= \dfrac{3.2.1}{2.1} \times \dfrac{6.5}{1.2} + \dfrac{3.2.1}{2.1} \times \dfrac{6}{1} + \dfrac{1}{1} \cdot \dfrac{1}{1}$

    $= 45 + 18 + 1$

    $= 64$

13. SIMPLE contains 6 different letters. There are two vowels I & E. Taking them as a single entity;

The letters to be arranged are SMPL(IE). There five 'letters' can be arranged in $\lfloor 5 = 5.4.3.2.1 = 120$ ways

The vowels (IE) may be arranged among themselves in $\lfloor 2 = 2$ ways

The number of words, each having the two vowels together = 120 × 2 = 240 → (i)

The total number of words formed by using all the 6 letters of the given word = $\lfloor 6 = 6.5.4.3.2.1 = 720$ (ii)

From (i) & (ii) the total number of words each having the two vowels never together = 720 - 240 = 480

14. The number should be divisible by 5. So the units place must be occupied by '0'. This is done in one way. There are six other numbers to select the thousands place since no digit should be repeated

This can be done in $6c_1 = 6$ ways. Having selected 0, & this chosen number, there are five other numbers from which one can be selected to occupy hundredths place. This is done in $5c_1 = 5$ ways lastly one of the other four can be chosen to occupy the tenth's place in $4c_1 = 4$ way

∴ The total no of ways of selection is 1 × 6 × 5 × 4 = 120 ways

15. S A N C T I O N has 8 letters (7 different letters)

1S 1A, N,N 1C, 1T, 1I, 1'O'. There '8' letters can be considered as 5 entities

If (AIO) are to occur together;

S (2N) C T (AIO). They can be arranged among themselves in
$$\frac{\lfloor 6}{\lfloor 2} = \frac{6.5.4.3.2.1}{1.2} = 360$$

The vowels can be arranged among themselves in $\lfloor 3 = 6$ ways

Required no $\left(\dfrac{\lfloor 6}{\lfloor 2}\right) \lfloor 3 = \dfrac{6.5.4.3.2.1}{1.2} 3.2.1 = (360)(6) = 2160$

16. R A̲ N K I̲ N G. Treat the 2 vowels as one entity

Among the others N repeats others are different.

No. of ways of arranging the letters

R N K N G (AI) in which N overs twice, others are different

$$= \frac{\lfloor 6}{\lfloor 1 \ \lfloor 2 \ \lfloor 1 \ \lfloor 1 \ \lfloor 1} = \frac{6.5.4.3.2.1}{2} = 360$$

The two vowels AI can be arranged among themselves in $\lfloor 2 = 2$ ways

Reqd no. of ways = $360 \times 2 = 720$

17. O P E R A T I O N we treat the 5 vowels O E A I O as one entity.

We then have P R T N (O E A I O) 5 entities. These five can be arranged in $\lfloor 5 = 5.4.3.2.1 = 120$ ways

The 5 vowels in which 'O' occurs twice, others being different can be arranged in $\dfrac{\lfloor 5}{\lfloor 2} = \dfrac{5.4.3.2.1}{1.2} = 60$ ways

Reqd. no of ways = $120 \times 60 = 7200$ ways

Key

1)	c	2)	b	3)	b	4)	d	5)	d	6)	c
7)	c	8)	d	9)	d	10)	d	11)	c	12)	b
13)	c	14)	b	15)	c	16)	c	17)	c		

36

Counting

Introduction

1. Using systematic listing (**Practice Exercises**)

    Ex.: 1 If how many distinct products can be formed with x & y?

    Sol.:
$1 \times 1 = 1$	$2 \times 1 = 2$	$3 \times 1 = 3$	$4 \times 1 = 4$
$1 \times 2 = 2$	$2 \times 2 = 4$	$3 \times 2 = 6$	$4 \times 2 = 8$
$1 \times 3 = 3$	$2 \times 3 = 6$	$3 \times 3 = 9$	$4 \times 3 = 12$
$1 \times 4 = 4$	$2 \times 4 = 8$	$3 \times 4 = 12$	$4 \times 4 = 16$
$1 \times 5 = 5$	$2 \times 5 = 10$	$3 \times 5 = 15$	$4 \times 5 = 20$

    The distinct products are 1, 2, 3, 4, 5, 6, 8, 9, 10, 12, 15, 16, 20

    13 in number; of course (a) prime numbers > 5; and (b) their multiples [(a) like = 7, 11, 13] [(b) like 14 (7 × 2); 17, 18 (2 × 3 × 3), 19] are not there

    Check 13 + 7 = 20 (7, 11, 13, 14, 17, 18, 19)

2. **Using multiplication Principle**

    (When listing is not practicable)
    If one event can happen in m ways, another in n ways, both events can happen in m × n ways

    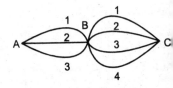

    Ex.: 2 If there are 3 roads from A to B & four from B to C there are $3 \times 4 = 12$ ways from A to C through B.

    Ex.: 3 Combination lock: (with 3 digits)

    Since any digit from 0 to 9 can be selected to occupy any position there are $10 \times 10 \times 10$ different 3 digit combinations possible for a combination lock.

Ex.: 4 Outfits: If there are 4 ties, 4 shirts & 4 trousers that a person possesses there will be $4 \times 4 \times 4 = 64$ different outfits he can wear.

Ex.: 5 Menu: If there are 3 different 'dhals', 5 different 'curries' 3 different 'chutnies' 4 different "sambhars/ rasas", & 2 variety of 'curds' there are $3 \times 5 \times 3 \times 4 \times 2 = 360$ varieties of food that a hotel can serve.

Ex.: 6 With 'repetition' & 'Without' repetition'

Using the numbers 1, 2, 3, 4 how many three digit numbers less than 250 can be formed (a) if the same number cannot be used more than once & (b) if the same digit can be used more than once.

☐ ☐ ☐
A B C

a) 12   b) 32   c) 44   d) 66

Sol.: We multiply the number of possible digits that can be used to fill the three blank positions A, B, C

☐ ☐ ☐
A B C

Since the three digit number formed should be < 250, there are only two possible choices for the first digit out of 1, 2, 3, 4; 1 or 2

In the Hundred's place denoted by [A]

With out repetition:-

Any one of the remaining three can be used to fill the second position There are now only two more numbers to choose one of them for the third position.

Using the multiplication principle of counting the 3 digit number can be formed in $2 \times 3 \times 2 = 12$ ways

With repetition:

First position can be filled by either 1 or 2. 2nd place by using any one of four digits 1, 2, 3, 4 third also by one of four so, here the no of ways is $2 \times 4 \times 4 = 32$ ways

Total $12 + 32 = 44$ numbers < 250

Check Listing out them
With out repetition

   123   213
   124   214
   132   231
   134   234
   142   241
   143   242

$6 \times 2 = 12$

With repetition

111	121	131	141	211	221	231	241
112	122	132	142	212	222	232	242
113	123	133	143	213	223	233	243
114	124	134	144	214	224	234	244

$4 \times 8 = 32$ Total = 44

Using venn Diagram

If n (A), n (B) denote the no of items in the sets A & B; then the no of items that belong to both the sets [denoted by n (A or B)]

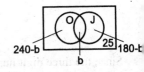

n (A or B) = n (A) + n (B) - n (A & B)

Ex.: 7 Of the 410 students in a college, 240 study oracle, 180 study Java, If 25 study neither, how many study both?

(a) 25   (b) 35   (c) 180   (d) 240   (e) 410

The no of students who study at least one language is 410 - 25 = 385

If 'b' study both; (240 - b) study only oracle; (180 - b) only Java; then

385 = (240 - b) + (b) + (180 - b) = 420 - b

∴   b = 420 - 385

     = 35

35 study both

Permutations: Arranging n objects in n slots.

Combination lock

Ex.:1 Using the integers 2, 5, 6 how many lock combinations can be formed without repetition?

(a) 3   (b) 4   (c) 6   (d) 12

Sol.: Any of the three no's can be filled in 1st position, Any of the other two: In the second, the third one in the 3rd position

$3 \times 2 \times 1 : 6$ ways

In general the number of different ways in which n different objects can be arranged in a line is $\underline{|n} = n(n-1)(n-2)\ldots 2:1$

Arranging n objects in a fewer seats

Ex.:2 If there are 8 students in a race, the no of different arrangements of the 1st, 2nd & 3rd places that are possible is $(8) \times (7) \times (6) = 336$

Ex.:3 Telephone numbers

NRI Gopal is back in India & remembers that his friend's telephone no is a 3 digit number in a village, all even, non repeating numbers. How many maximum no of telephone calls he has to make before he dials the correct number?

(a) 4   (b) 12   (c) 24   (d) 48   (e) 96

Sol.: There are five even numbers 0, 2, 4, 6, 8. Since the first number can not be zero the first place can be filled by any one of the other four. 2nd place can be filled by any number, not already used, including zero. There are four choices. For the third place there will be 3 digits possible

So the no of calls, maximum he has to make is $4 \times 4 \times 3 = 48$ if he (unfortunately) succeeds only in the last attempt!

Combinations

A combination is an unordered set of objects. Permutation is any ordered arrangement of the objects in that set

Ex.: 4 Take the first 3 prime numbers 2, 3, 5. There is only one combination of them, but there are 6 permutations: They are

| 235 | 325 | 523 |
| 253 | 352 | 532 |

Quantitative Aptitude

Combination formulae:-

The number of different subsets that contain r elements of a parent set consisting of n objects is nc_r (with out regard to their order)

Read as the no of combinations of n objects taken r at a time. It is

$$n_{c_r} = \frac{\lfloor n}{\lfloor r \; \lfloor n-r}$$

Ex.: 5 The no of different combinations of the numbers taken two at a time (r = 2) from a set of three numbers 1, 2, 3 (n = 3) are nc_r

Hence $3c_2 = \dfrac{\lfloor 3}{\lfloor 2 \; \lfloor 1} = \dfrac{3.2.1}{1.2.1} = 3$

Namely (1, 2), (1, 3), (2, 3)

These are the 3 combinations (r = 2, n = 3) from a set of 3 numbers, taken two at a time.

Ex.: 6 How many different combinations (committees) of strength 3 can be formed from a set of 6 persons of different religions?

(a) 9   (b) 12   (c) 15   (d) 20

Sol.: $6c_3 = \dfrac{\lfloor 6}{\lfloor 3 \; \lfloor 3}$

$= \dfrac{6.5.4}{1.2.3} = 20$

Ex.: 7 Buffet party in an evening snack party, 3 savories, 3 sweets & two drinks are prepared. What is the total number of different combinations consisting of one sweet, one savory, & one drink?

(a) 6   (b) 12   (c) 15   (d) 18

Sol.: $3c_1 \; 3c_1 \; 2c_1 = \dfrac{\lfloor 3}{\lfloor 1 \; \lfloor 2} \; \dfrac{\lfloor 3}{\lfloor 1 \; \lfloor 2} \; \dfrac{\lfloor 2}{\lfloor 1 \; \lfloor 1} = 3 \times 3 \times 2 = 18$

Permutations subject to conditions

Ex.: 8 There are five varieties of flowers in a tray. Roses, Jasmines, Tulips, Daffodils & lilies. How many different combinations of three flowers keeping always rose as the first, can be formed?

(a) 4   (b) 6   (c) 8   (d) 12   (e) 24

Sol.: ☐ ☐ ☐

   1   2   3   Rose in 1st position, 2nd place can be any one of the remaining four, third any one of the remaining three.

∴ The no of combinations is $1 \times 4c_1 \times 3c_1$

$$= 1 \times \frac{\lfloor 4}{\lfloor 1 \lfloor 3} \cdot \frac{\lfloor 3}{\lfloor 1 \lfloor 2} = 1 \times 4 \times 3 = 12$$

Explicitly writing them; they are

1. R J T	5. R T D	9. R D L
2. R J D	6. R T L	10. R L J
3. R J L	7. R D J	11. R L T
4. R T J	8. R D T	12. R L D

Ex.:9 In how many different ways can four students be seated in two chairs?

(a) 4   (b) 6   (c) 8   (d) 12

Sol.: Any one of the four can sit in the first. Any one of the other three then in the second.

∴ $4 \times 3 = 12$ ways

Ex.: 10 In a class of 40, 16 students participate in games, 13 in cultural activities, eight take part in both. How many are neither participating in games nor in cultural actuaries?

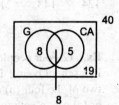

(a) 8   (b) 13   (c) 16   (d) 19

Sol.: Draw the venn diagram. Students participating in games alone 8; in C. A' alone = 5; In both = 8; Remaining = 40 - (8 + 8 + 5)
In none = 19

Ex.: 11 Permutations with Conditions:

How many three digit numbers greater than 100 can be formed from the 5 digits 0, 1, 2, 3, 4 if the same digit cannot be used more than once?

(a) 24   (b) 48   (c) 72   (d) 100

Sol.: ☐ ☐ ☐

We cannot use 0 as the first digit. Hence there are 4 choices for the first blank. The same no cannot be repeated; so we have four choices for the second position; Three choices for the third;

Hence $4c_1 \times 4c_1 \times 3c_1 = \dfrac{\lfloor 4}{\lfloor 1 \ \lfloor 3} \dfrac{\lfloor 4}{\lfloor 1 \ \lfloor 3} \dfrac{\lfloor 3}{\lfloor 1 \ \lfloor 2}$

$= 4 \times 4 \times 3$

$= 48$ ways to form 3 digit nos.

Greater than 100; writing explicitly, these are

Check

102	130	201	230	301	320	401	420
103	132	203	231	302	321	402	421
104	134	204	234	304	324	403	423
120	140	210	240	310	340	410	430
123	142	213	241	312	341	412	431
124	143	214	243	314	342	413	432

48

Ex.: 12
Four people are contesting in an election to fill two vacancies in a school board. In the election, a voter (a) may cast a vote for any two of the four (b) cast a vote for exactly one of the four candidates, or (c) cast a vote for none of the four candidates. What is total no of different choices a voter has when casting a vote?

(a) 9   (b) 11   (c) 16   (d) 17

Sol.: If voter may cast a vote for two of the four candidates. In $4c_1 \times 3c_1$ i.e 12 ways

A voter may cast a vote for exactly one of the four candidates in $4c1$ i.e 4 ways $\dfrac{\lfloor 4}{\lfloor 1 \, \lfloor 4-1} = 4$

A voter may vote for none of the four in one way $(4_{c_0}) = \dfrac{\lfloor 4}{\lfloor 4-0} = 1$

Total no of different voting choice is $12 + 4 + 1 = 17$

Ex.: 13 In a university the number of persons registered in the engineering stream, management stream, Biology stream are mentioned in the venn diagram. What is the approximate percentage of students who are interested in all the three streams?

(a) 2.567   (b) 3.892   (c) 4.347   (d) 5.2131

Total in Engg. stream = 48

Total in Mgt stream = 38

Total in Biology stream = 46

Common to all is 4 out of a total of

$24 + 6 + 16 + 14 + 4 + 12 + 16 = 92$

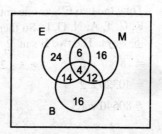

Percentage $= \dfrac{4}{92} = 4.347\%$

Ex.: 14 The telephone numbers in a village start with the number 2 (code) the next three digits should neither start with nor end with a zero. How many numbers may be assigned?

(a) 420   (b) 540   (c) 647   (d) 810

Sol.: ☐ ☐ ☐ ☐
       2   not   not
             Zero   Zero

The second place i.e 100 place can be filled with any no from 1 to 9

The third place with any number from 0 to 9; Fourth place with any number from 1 to 9;

Slots 1   2   3   4

∴ $1 \times 9 \times 10 \times 9 = 810$ numbers

Ex.: 15 Six points lie in a plane (no three in a line) A circle can be drawn through any three points in a plane. How many circles can be drawn?

(a) 13   (b) 15   (c) 18   (d) 20

$$6c_3 = \frac{\underline{|6}}{\underline{|3}\,\underline{|3}} = \frac{6.5.4}{1.2.3} = 20$$

Ex.: 16 In how many different ways can the individual letters of the word RECTANGLE be arranged in a line such that the two E'S appear consecutively?

(a) 720   (b) 5040   (c) 40320   (d) 80640

Treat the two E'S as one entity. There are seven distinct other letter its R, C, T, A, N, G, L. So there are 8 'letters'. They can be arranged in $\underline{|8}$ ways. The two E's in $\underline{|2}$ ways among themselves

$= (8 \times 7 \times 6 \times 5 \times 4 \times 3 \times 2 \times 1) \times (1 \times 2)$

$= 40320 \times 2$

$= 80640$ ways

37

Probability

Exercises

1. X attends the office 80% of the time & Y 90% of the time. What is the percentage of days when either of them is absent?

    (a) 17%   (b) 22%   (c) 26%   (d) 30%

2. What is the probability of getting a sum of 8 in a simultaneous throw of a pair of dice?

    (a) $\dfrac{1}{18}$ (b) $\dfrac{5}{36}$ (c) $\dfrac{1}{6}$ (d) $\dfrac{1}{4}$

3. Three cards are drawn at random from a pack. What is the probability of getting one 'diamond', one 'club' & one 'heart'?

    (a) $\dfrac{13}{425}$ (b) $\dfrac{169}{425}$ (c) $\dfrac{169}{850}$ (d) $\dfrac{169}{1700}$

4. There are 3 red, 3 green, & 3 blue balls in a box. Three balls are drawn at random. What is the probability that none of them is blue?

    (a) $\dfrac{2}{21}$ (b) $\dfrac{1}{7}$ (c) $\dfrac{4}{21}$ (d) $\dfrac{5}{21}$

5. A box contains 5 red, 5 green & 5 white balls. What is the probability of drawing three balls of the same colour?

    (a) $\dfrac{5}{91}$ (b) $\dfrac{6}{91}$ (c) $\dfrac{7}{91}$ or $\dfrac{1}{13}$ (d) $\dfrac{8}{91}$

6. In a simultaneous throw of two coins, the probability of getting two heads is

   (a) 1/4   (b) 1/2   (c) 3/4   (d) 1

7. The probability of getting at least two heads is

   (a) 1/8   (b) 1/4   (c) 3/8   (d) 1/2

8. Four unbiased coins are tossed. What is the probability of getting at least two heads?

   (a) $\dfrac{5}{8}$ (b) $\dfrac{11}{16}$ (c) $\dfrac{3}{4}$ (d) $\dfrac{15}{16}$

9. In a throw of a die, what is the probability of getting a number greater than 1?

   (a) $\dfrac{1}{6}$ (b) $\dfrac{1}{3}$ (c) $\dfrac{5}{6}$ (d) 1

10. Two dice are thrown. What is the probability of getting a total of at least 8

    (a) 1/3   (b) 5/12   (c) 1/2   (d) 7/12

11. Two dice are thrown. What is the probability of getting two numbers whose product is odd?

    (a) 1/4   (b) 1/2   (c) 3/4   (d) 1

12. Two cards are drawn from a pack at random. What is the probability that either both are black or both are aces?

    (a) $\dfrac{49}{221}$ (b) $\dfrac{51}{221}$ (c) $\dfrac{53}{221}$ (d) $\dfrac{55}{221}$

13. From a group of 10 boys & 8 girls, four students are selected. What is the probability that two are boys & two are girls?

    (a) 6/17   (b) 7/17   (c) 8/17   (d) 9/17

14. A box contains 30 electric bulbs out of which 5 are defective. Three are chosen at random. The probability that at least one of them is defective is?

    (a) $\dfrac{55}{203}$ (b) $\dfrac{66}{203}$ (c) $\dfrac{77}{203}$ (d) $\dfrac{88}{203}$

15. Two dice are thrown. What is the probability that the sum of the numbers is a prime number?

    (a) 1/4   (b) 1/3   (c) 5/12   (d) 1/2

16. If the letters of the word ANGEL are randomly arranged to form 5 letter words, what is the probability that the result is ANGLE?

    (a) 1/120   (b) 1/60   (c) 1/40   (d) 1/20

17. Three fair coins are tossed, what is the probability that all will come up with heads or all with tails?

    (a) 1/8   (c) 1/2
    (b) 1/4   (d) 1

18. If two fair dice are tossed, what is the probability that both will show the same no?

    (a) 0   (b) $\dfrac{1}{2}$ (c) $\dfrac{1}{4}$ (d) $\dfrac{1}{6}$

Solutions

1. The probability that X attends the office

$$P(X) = 80\% = \frac{4}{5}$$

$$P(Y) = 90\% = \frac{9}{10}$$

Then

$$P(\bar{X}) = \frac{1}{5};$$

$$P(\bar{Y}) = \frac{1}{10}$$

The probability that one of them is absent is

P (X present & y absent or y present & x absent)

$$= P(X \& \bar{Y} \text{ or } \bar{X} \& Y) = P(X \& \bar{Y}) + \bar{P}(\bar{X} \& Y)$$

$$= P(X)P(\bar{Y}) + P(\bar{X})P(Y) = \frac{4}{5} \cdot \frac{1}{10} + \frac{1}{5} \cdot \frac{9}{10}$$

$$= \frac{4+9}{50} = \frac{13}{50} = \frac{26}{100}$$

(or) 26% (or) 0.26

2. If E is the event of getting a sum of 8; Then n (E) = ? (2, 6), (3, 5), (4, 4) (5, 3) & (6, 2)

$$\therefore n(E) = 5; n(s) = 36$$

$$\therefore p(8) = \frac{5}{36}$$

3. If S is the sample space of taking any 3 cards from a pack;

$$n(s) = 52c_3 = \frac{\lfloor 52}{\lfloor 3 \lfloor 49} = \frac{52.51.50}{1.2.3} = 1300 \times 17$$

If E is the event of getting 1 club, 1 diamond & one heart;

$$n(E) = 13c_1 \times 13c_1 \times 13c_1 = \frac{\lfloor 13}{\lfloor 1 \lfloor 12} \frac{\lfloor 13}{\lfloor 1 \lfloor 12} \frac{\lfloor 13}{\lfloor 1 \lfloor 12} = 13.13.13$$

$$\therefore P(E) = \frac{13.13.13}{1300 \times 17} = \frac{169}{1700}$$

4. If s is the sample space of drawing any three out of 9 balls.

$$n(S) = 9c_3 = \frac{\lfloor 9}{\lfloor 3 \lfloor 6} = \frac{9.8.7}{1.2.3} = 84$$

Three non blue balls from 6 non blue balls

$$n(E) = 6c_3 = \frac{\lfloor 6}{\lfloor 3 \lfloor 3} = \frac{6.5.4}{1.2.3} = 20$$

$$\therefore P = \frac{n(E)}{n(S)} = \frac{20}{84} = \frac{5}{21}$$

5. n(s) is sample space consists of 5r, 5w, 5G balls. Drawing any 3 balls from these: n(s) = 15_{c_3}

The 'Event' is they are of the same colour either all 3 red, or all 3 green or all 3 white.

$n(E) = 3 \times 5_{c_3}$

$$P = \frac{3 \times 5_{c_3}}{15c_3} = \left[\frac{3 \times \frac{\lfloor 5}{\lfloor 3 \ \lfloor 2}}{\frac{\lfloor 15}{\lfloor 3 \ \lfloor 12}}\right] = \left[\frac{3 \times \frac{5.4}{1.2}}{\frac{15.14.13}{1.2.3}}\right] = \frac{30}{5 \times 91} = \frac{30}{455} = \frac{6}{91}$$

6. HH   TH
    HT   TT
       1/4

7. Possible out comes 8; Desired HHH, HTH from HHT, THH
   $\frac{4}{8} = \frac{1}{2}$.

8. Sample space:

1) HHHH	2) HHHT	3) HHTH	4) HHTT
5) HTHH	6) HTHT	7) THHH	8) THHT
9) TTHH	10) TTHT	11) TTTH	12) TTTT
13) HTTH	14) HTTT	15) THTH	16) THTT

   Desired out come (16 - those underlined 5)

   [10, 11, 12, 14, 16]

   Elements out of 16; $\frac{11}{16}$

9. Out of six possible out comes (2, 3, 4, 5, 6): $\frac{5}{6}$

10. Out of sample space 36

    Event E is in the cases (2,6) (3,5) (4,4) (5,3) (6,2) / (3,6) (4,5) (5,4) (6,3) / (4,6) (5,5) (6,4) / (5,6) (6,5) / (6,6)

    15 cases P (success) ie P (18) = $\frac{15}{36}$ or $\frac{5}{12}$

11. Sample space S: n(S) = 36

    Desired outcomes shown as (✓) (product being odd): as shown below

11✓	21	31✓	41	51✓	61
12	22	32	42	52	62
13✓	23	33✓	43	53✓	63
14	24	34	44	54	64
15✓	25	35✓	45	55✓	65
16	26	36	46	56	66

The success (product being odd) is ticked (✓)

These are in 9 situations

$$\therefore \text{ Prob required} = \left(\frac{9}{36}\right) = \frac{1}{4}$$

12. Sample space 52

    Picking any two from the pack is $n(s) = 52c_2$

    $$= \frac{52.51}{1.2} = 26.51 = 1326$$

    Events are

    (i) both are blacks; $n(E_1) = 26c_2$

    $$\frac{26.25}{1.2} = 325;$$

    $$P(E_1) = \frac{325}{1326}$$

    (ii) both are aces $n(E_2) = 4c_2$

    $$= \frac{4.3}{1.2} \quad P(E_2) = \frac{6}{1326} = 6$$

    (iii) 2 aces (one of clubs & one of spades) this event is repeated in i & ii

    $$n(E_1 \cap E_2) = 2c_2$$

    $$= \frac{\lfloor 2}{\lfloor 2 \lfloor 0} = 1; \text{ Its probability is}$$

    i.e. $P(E_1 \cap E_2) = \frac{1}{1326}$

    $$\therefore P(E_1 \cup E_2) = P(E_1) + P(E_2) - P(E_1 \cap E_2)$$

    $$= \frac{325 + 6 - 1}{1326} = \frac{330}{1326} = \frac{55}{221}$$

13. $n(s) = 18c_4 = \dfrac{18.17.16.15}{1.2.3.4} = (51.60)$

$n(E_1) = 10c_2 = \dfrac{10.9}{1.2} = 45$

$n(E_2) = \dfrac{8.7}{1.2} = 28$

We have to select 2 boys & 2 girls i.e $n(E_1) \times n(E_2)$

This $= 45 \times 28$

$\therefore$ Required prob $= \dfrac{n(E_1)\,n(E_2)}{n(s)} = \dfrac{45 \times 28}{51 \times 60} = \dfrac{7}{17}$

14. Possible out comes

$n(s) = 30c_3 = \dfrac{30.29.28}{1.2.3} = 5 \times 29 \times 28$

P(E) = P (at least one is defective) = [1 - p (none is defective)]

P (none is defective) = ?, if this event' is success

$n(E) = 25c_3$

$\therefore$ P (none is defective)

$\left(\dfrac{\frac{25.24.23}{1.2.3}}{5 \times 29 \times 28} \right) = \dfrac{25 \times 4 \times 23}{5 \times 29 \times 28} = \dfrac{115}{203}$

$\therefore$ P(at least one is defective) $= 1 - \dfrac{115}{203} = \dfrac{88}{203}$

15. Possible out comes; Desired out come n(E)

N(s) = 36; n(E) =?

E:
(1,1)	(2,1)	(3,2)	(4,1)	(5,2)	(6,1)
(1,2)	(2,3)	(3,4)	(4,3)	(5,6)	(6,5)
(1,4)	(2,5)				
(1,6)					

$\therefore$ n(E) = 15

$\therefore$ P $= \dfrac{15}{36} = \dfrac{5}{12}$

16. The probability that A is selected for the first blank space is 1/5

    Problem that N is selected for the second from among the other four i.e N, G, E, L is 1/4th etc

    $\therefore$ Problem of forming the word 'ANGLE'

    $= \dfrac{1}{5} \times \dfrac{1}{4} \times \dfrac{1}{3} \times \dfrac{1}{2} \times \dfrac{1}{1} = 1/120$

17. HHH
    TTT
    $\therefore$ Probability is $\dfrac{1}{2} \times \dfrac{1}{2} \times \dfrac{1}{2}$;

    $\therefore$ Probability is $\dfrac{1}{2} \times \dfrac{1}{2} \times \dfrac{1}{2}$

    Probability that all will be heads or all will be tails

    Total $= \dfrac{1}{8} + \dfrac{1}{8} = \dfrac{1}{4}$

18. $n(s) = 6 \times 6 = 36;$

    E = (1, 1), (2, 2), (3, 3), (4, 4), (5, 5), (6, 6)

    $\therefore n(E) = 6$

    $\therefore P(E) = \dfrac{6}{36} = \dfrac{1}{6}$

Key

1)	c	2)	b	3)	d	4)	d	5)	b	6)	a
7)	d	8)	b	9)	c	10)	b	11)	a	12)	d
13)	b	14)	d	15)	c	16)	a	17)	b	18)	d

38

Data Interpretation - Tabulation

Introduction

Lengthy data is some times condensed to tables, bar graphs, line graphs, pie charts etc. Interpretation of such condensed data, analysis of the data, finding whether, the data is 'sufficient' finding which data is 'necessary' & which data is 'redundant' 'Yes' or 'No' questions (on statements providing 'sufficient' information or not) are considered. + ' word problems' are another variation of testing the candidates' skills.

Solved Exercises

1. The table gives the percentage distribution of population of five states A, B, C, D, E, literacy and sex. Six sets of Questions follow:

Table 1

State	Percentage of illiterate population	Population of Males & Females	
		Illiterate M : F	Literate M : F
A	90	5 : 10	9 : 6
B	80	2 : 3	5 : 3
C	60	1 : 3	7 : 3
D	30	1 : 4	3 : 2
E	20	1 : 4	2 : 1

(a) Population of state A is 6 million. What is the number of females who are literate?
    (a) 0.15 millions   (b) 0.24 millions   (c) 1.8 millions
    (d) 2.4 millions   (e) 3.2 millions   (f) 3.6 millions

(b) If the population who are literate is 4 millions in state B, its total population is

    (a) 5 millions   (c) 12 millions   (e) 20 millions
    (b) 10 millions   (d) 18 millions

(c) If the female population who are illiterate is 9 million. What is the male population who are literate?

    (a) 6.3 millions   (c) 18.9 millions
    (b) 12.6 millions   (d) 25.2 millions

(d) If the illiterate population for state D is 3.6 millions and that for state E is 5 millions then the total populations of states D and E are in the ratio

    (a) 5 : 20   (b) 8 : 20   (c) 6 : 25   (d) 9 : 25   (e) 12 : 25

Solutions

1. (a) Population of state A is 6 million; Illiterate population in state A is 90%.

$$= \frac{90}{100} \times 6 \text{ million} = 5.4 \text{ millions}$$

Literate population $= 6 - 5.4 = 0.6$ millions

$$\text{Illiterate Males} = \frac{5}{15} \times 5.4 \text{ millions}$$

$$= \frac{1}{3} \times 5.4 \text{ millions}$$

$$= 1.8 \text{ millions}$$

$$\text{Illiterate Females} = \frac{10}{15} \times 5.4 \text{ millions}$$

$$= \frac{2}{3} \times 5.4 \text{ millions}$$

$$= 3.6 \text{ millions}$$

Data Interpretation - Tabulation 285

Literate Males $= \dfrac{9}{15} \times 0.6$ millions

$= \dfrac{3}{5} \times 0.6$ millions

$= 0.36$ millions

Literate Females $= \dfrac{6}{15} \times 0.6 = \dfrac{2}{5} \times 0.6$ millions

$\dfrac{1.2}{5}$ millions

$= 0.24$ millions

(in millions)

State A Summary	Illiterate		Literate		Total	
	M	F	M	F	M	F
	1.8	3.6	0.36	0.24	1.80	3.60
Total Literate	–		0.60		0.36	0.24
Total Illiterate	5.4		–			
Population	–		–		2.16	3.84
TOTAL					6.00	6.00

Total Illiterate $= 5.4$ millions

Total Literate $= 0.60$ millions

Total Population $= 6$ millions

(b) Literate population in state B $= 4$ millions

Since Illiterate population is 80% in B,

Literate population is 20% in B

If 20 is Literate $\rightarrow$ Total population is 100

So, if 4 million are literate

Total population of B is $\dfrac{100}{20} \times 4 = 20$ millions

Other particulars (not asked)

Illiterate population in B is 16 millions

Literate Males $= \dfrac{5}{8} \times 4 = 2.5$ millions

Literate Females $= \dfrac{3}{8} \times 4 = 1.5 \text{ millions}$

Illiterate Males in B $= \dfrac{2}{5} \times 16 = \dfrac{32}{5} = 6.4$ millions

Illiterate Females in B $= \dfrac{3}{5} \times 16 = \dfrac{48}{5} = 9.6$ millions

Summary of B (In mill)	Males		Females		Total M	Total F
	Lit.	Illiter.	Lit.	Illiter.	8.9	11.1
	2.5	6.4	1.5	9.6		
Total Literate	2.5	–	1.5	–	4	–
Total Illiterate	–	6.4	–	9.6		16

Total population → 20 millions

(c) Let the total population of state C be P.

Literate population = 60% or 0.6p

Illiterate population = 0.4 p

Literate Males $= \dfrac{7}{10} \times$ Literate population

$= \dfrac{7}{10} \times 0.6p$

$\dfrac{4.2}{10}p = 0.42$ p m;

Literate F $= \dfrac{3}{10} \times$ Literate population

$= \dfrac{3}{10} \times 0.6p$

$= \dfrac{1.8}{10}p$

$= 0.18$ population

Illiterate M $= \dfrac{1}{4} \times$ Illiterate population

$= \dfrac{1}{4} \times 0.4p$

Data Interpretation - Tabulation

$$= 0.1\,p$$

Illiterate F $= \dfrac{3}{4} \times$ Illi.population

$$= \dfrac{3}{4} \times 0.4p$$

$$= 0.3p$$

Now, given $=$ Ill. Females $= 9$ millions

$\therefore\ 0.3\,p = 9$ millions

Or p $= \dfrac{9 \times 10}{3} = 30m$

Then Liter. M $= 0.42 \times 30 = 12.60$ millions

Summary of C (in mill)

Population	Males		Females	
	Lit.	Illiter.	Lit.	Illiter.
30 millions	0.42 p	0.1 p	0. 18 p	0.3 p
	12.60	3	5.4	9

Illiterate Males $= 0.1\,p$

$$= 0.1 \times 30$$

$$= 3\ \text{millions}$$

Literate Females $= 0.18\,p$

$$= 0.18 \times 30$$

$$= 5.40\ \text{millions}$$

Literate Males $= 0.42p = 0.42 \times 30 = 12.60$ millions

Illiterate Females $= 9$ millions

(d) Illiterate population in D $= 3.6$ millions

(Lit: Illiterate) population in D is 70:30

Illiterate population in E $= 5.0$ millions If L_D is the lit. prop in D

(Lit: Illiterate population) in E is 80:20 and L_E is the lit. prop in E;

Since L_D: 3.6 millions $= 7 : 3$

Then $70 \times 3.6 = 30 L_D$

$\therefore L_D = \dfrac{70 \times 3.6}{30} = 8.4$ millions

Similarly $L_E : 5M = 80:20$

$20 L_E = 80 \times 5$ millions

$L_E = \dfrac{80 \times 5}{20} = 20$ millions

Now (Ill M: Ill F) in D is $1:4$; $\dfrac{1}{5}$ of ill P is Ill m and $\dfrac{4}{5}$ Ill p is Ill F

$\dfrac{1}{5} \times 3.6 = 0.72 m$ is Ill m in D

And $\dfrac{4}{5} \times 3.6 = \dfrac{14.4}{5} = 2.88$ M is Ill F in D

Literate $M = \dfrac{3}{5}$ Literate population of D

$= \dfrac{3}{5} \times 8.4 = 5.04$ millions

Literate $F = \dfrac{2}{5}$ Literate population if $D = \dfrac{2}{5} \times 8.4 = 3.36$ millions

Check In D Lit M : Lit F = 5.04 : 3.36

Is it = 3 : 2?

$2 \times 5.04 = 1008$

$3 \times 3.36 = 10.08$; Is Ill M: Ill F in D

$0.72 : 2.88 = 1 : 4$? yes

Is L : I population in D

i.e., $8.4 : 3.6 = 7 : 3$? yes QED

Summary of D	Males	Females	Total
Literates	5.04 m	3.36 m	8.4m
Illiterates	0.72m	2.88m	3.6
Total	5.76m	6.24m	12

State E

Ill population in E = 5 millions; Lit p = 20m

(Ill: Lit) = 20:80 = 1:4; Total p = 25 m

Ill M = $\frac{1}{5}$ of Ill population of E = 1m

Ill F = $\frac{4}{5}$ of Ill populate E = 4m

Lit Males = $\frac{2}{3}$ of Lit population in E

Lit F = $\frac{1}{3}$ of Lit population in E

Lit population in E = 20 millions

Total population = 25 millions

Ill M = $\frac{1}{5} \times 5$ = 1 million

Ill F = $\frac{4}{5} \times 5$ = 4 millions

Lit Males = $\frac{2}{3} \times 20 = \frac{40}{3}$ = 13.33 millions

Lit F = $\frac{1}{3} \times 20$ = 6.67 millions

Summary of E (in millions)

	Males	Females	Total
Lit	13.33	6.67	20millions
Ill	1	4	5millions
Total	14.33	10.67	25millions

Quantitative Aptitude

Prob 1: Percentage marks of 100 students are given in the table. Answer the following Questions

% Marks → Subject ↓	≥ 90	≥ 70	≥ 60	≥ 50	> 40	≥ 20	≥ 0
Maths	20	40	50	70	80	90	100
Physics & Chemistry	5	10	35	50	65	80	100
Biology	3	8	20	30	40	60	100

Criteria for selection of students in the various courses

Category	Discipline	M	P & C	B
General	(A) Certain Engg. courses	≥ 70%	≥ 60%	
General	(B) Certain Biology oriented Technological courses & Medical courses		≥ 60%	≥ 60%
General	(C) M.Sc Physics & Chemistry courses		≥ 70%	
General	(D) M.Sc Biology			≥ 60%
Special Categories	(E) Certain Engineering courses	≥ 60%	≥ 40%	
Special Categories	(F) Medical & Biologes oriented Technical course		≥ 40%	≥ 50%
Course fee waiver	(G) Certain Engineering courses	≥ 90%	≥ 70	
Course fee waiver	(H) Medical & Biology oriented courses		≥ 70	≥ 70
"	(K) M.Sc Physics & chemistry courses		≥ 90	
"	(L) M.Sc Biology course			≥ 60

Solutions

(a) How many are qualified for "Certain Engineering courses"? (A)

(b) for B?   (e) for C?   (h) for H?
(c) for D?   (f) for F?   (i) for K?
(d) for E?   (g) for G?   (j) for L?

	M	P & C	B	Eligible students in percentage (i.e. out of hundred)
8(A)	≥70%(40)	≥60%(35)		(35)
8(B)		≥60%(35)	≥60%(20)	(20)
(8C)		≥%(70)		(10)
(8D)			≥60%	(20)
8(E)	≥60%(50)	≥40%(65)		(50)
8(F)		≥40%(65)	≥50%(30)	(30)
8(G)	≥90%(20)	≥70%(0)		(10)
8(H)		≥70%(10)	≥70%(8)	(8)
8(K)	≥90%(5)			(5)
8(L)			≥60%(20)	(20)

"Certain Engineering Courses" 35 + 50	Biology oriented Technology 20 + 30	M.Sc P&C 10	M.Sc B 20	Total 165+
Course waiver 10	8	5	20	43

165 as against total 100 **candidates**

39

Bar Graphs

Data is shown as vertical or horizontal bars. The lengths or (heights) and sometimes their areas) give the magnitude depending on the scale of the graphs. Questions asked are to be answered after analyzing the given data

Q.1 The income and expenditure (in lakhs of rupees) in the year 2005 of different companies are shown in bar graphs

Note : Percent profit or loss is calculated as = $\left[\dfrac{\text{Income - Exp}}{\text{Exp}} \times 100 \right]$

Answer the Questions

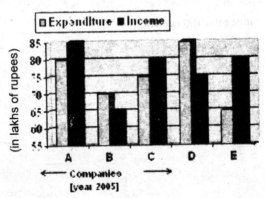

Questions:

1.A What is the percentage profit/loss of different companies in the year 2005?

1.B What is the percent profit / loss of all the companies together in 2005?

1.C Which company earned max. percentage profit in 2005? Which the least?

1.D What is the profit/loss of companies A and B together?

1.E If the income of C in the year 2005 was 10% more than its income in 2004 and it has earned a profit of 10% in 2004, its expenditure in 2004 in lakhs of rupees was?

1.F If for the company D the expenditure had increased by 10% in 2005 from year 2004, and if it had earned a profit of 10% in 2004, its income in 2004 in lakhs of rupees?

1.G Which company had the least loss in 2005? Which the max? How much?

(Lakhs of Rs.)

Company	Expenditure	Income
A	80	85
B	70	65
C	75	80
D	85	75
E	65	80
Total → lakhs	(375)	(385)

Solutions

1.A $\dfrac{\text{percent profit }(+)}{\text{or loss }(-)} = \dfrac{\text{Income - Exp}}{\text{Exp}} \times 100$

A $\dfrac{85-80}{80} \times 100 = \dfrac{5}{80} \times 100 = \dfrac{25}{4} = 6.25\%$

B $\dfrac{65-70}{70} \times 100 = \dfrac{-5}{70} \times 100 = -7.143\%$

C $\dfrac{80-75}{75} \times 100 = \dfrac{20}{3} = +6.666\%$

D $\dfrac{75-85}{85} \times 100 = \dfrac{-200}{17} = -11.7647\%$

E $\dfrac{80-65}{65} \times 100 = \dfrac{300}{13} = 23.0769\%$

1.B Total income of all companies in 2005 - Expenditure $\Big\} = 385 \text{ lakhs} - 375 \text{ lakh}$

Net profit $= \dfrac{385-375}{375} \times 100 = \dfrac{8}{3} = 2.6666$

1.C Max. profit E = 23.0769%
Least profit A = 6.25%

1.D Total income of A and B = 150 lakhs
Total Exp of A and B = 150 lakhs
Profit of (A & B) = 0%

1.E Suppose income of C in 2004 is I lakh

Income of C in 2005 = $\dfrac{110}{100}$ I lakhs

= 80 lakh (given data)

$\therefore$ I = $\dfrac{80 \times 100}{110}$ = 72.7273

Suppose expenditure of C in 2004 is E lakh profit in 2004 is 10% (given)

$$= \dfrac{I - E}{E} \times 100 = \dfrac{72.7273 - E}{E} \times 100$$

$\therefore \dfrac{72.7273}{E} - 1 = \dfrac{1}{10}$

$\therefore \dfrac{72.7273}{E} = \dfrac{11}{10}$

$\therefore E = \dfrac{72.7273 \times 10}{11} = 66.1157$

Required = 66.1157 lakh

1.F Let the expenditure of D in 2004 be E_D.

Exd in 2005 of D is $\dfrac{110}{100} E_D$ = 85 (given)

$\therefore E_D = \dfrac{100 \times 85}{110} = \dfrac{850}{11} = 77.2727$ lakh

Let the Income of D in 2004 be I_D.

The profit / loss = $\dfrac{I_D - 77.2727}{77.2727} \times 100 = 10$

Then $\dfrac{I_D}{77.2727} - 1 = \dfrac{1}{10}$ or $\dfrac{I_D}{77.2727} = \dfrac{11}{10}$

$\therefore I_D = \dfrac{11}{10} \times 77.2727 = 84.9999$ lakh ≈ 85

In 2004	In 2005
I E 85 77.2727 Profit = 10% Exp in 2005	I E 75 85 Loss = $\dfrac{75-85}{85} \times 100 = 11.7647\%$

Required = 84.99999 lakhs $\simeq$ 85 lakhs (Income in 2004)

1.G   Least loss in 2005 = B = − 7.143%

      Max. Loss in 2005 = D = − 11.7647%

1.H   What is the ratio of highest profit to highest loss in 2005?

$$\dfrac{23.0769}{11.7647} = 1.9615$$

1.I   What is the ratio of highest profit to Average profit of all companies in 2005?

$$\dfrac{\text{Highest profit}}{\text{Average profit of all companies}} = ?$$

$$= \dfrac{23.0769}{2.6666} = 8.6541$$

40

Pie Charts

1.

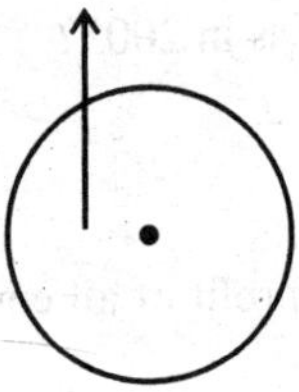

Fig.1

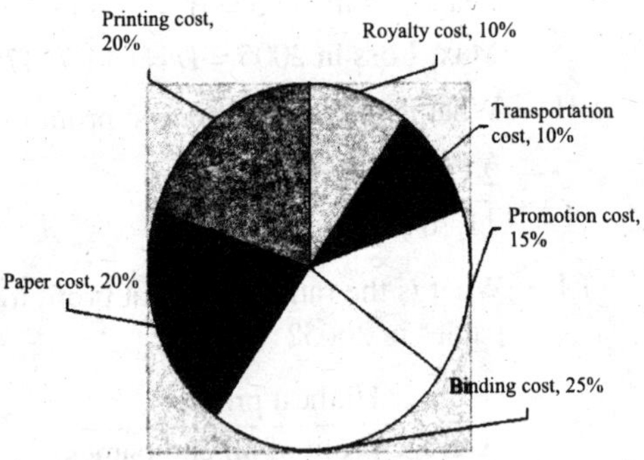

Fig.2 (Solution)

Printing cost, 20%
Royalty cost, 10%
Transportation cost, 10%
Promotion cost, 15%
Binding cost, 25%
Paper cost, 20%

The percentage distribution of the expenses in the publication of a book are shown in the pie chart (to be prepared) prepare the pie chart as per the details

Central angles

(a) printing cost 20% → $\dfrac{20}{100} \times 360 = 72°$

Paper cost 20% → $\dfrac{20}{100} \times 360 = 72°$

Binding cost 25% → $\dfrac{25}{100} \times 360 = 90°$

Promotion cost 15% → $\dfrac{15}{100} \times 360 = 54°$

Transportation cost 10% → $\dfrac{10}{100} \times 360 = 36°$

Royalty cost 10% → $\dfrac{10}{100} \times 360 = 36°$

Total = 100%   Total = 360°

Prepare the pie chart starting with the vertical radius (fig. 1)
Solution : fig. 2

(b) If for an edition of the book the paper cost is Rs.75000/-, what are the different costs?

Sol: 20% is Rs.75000/- ∴ 100% is Rs.3,75,000/-

Printing cost = 20%   75,000

Paper cost = 20%   75000 (given)

Binding cost = 25% $\frac{25}{100} \times 3,75,000 = 93,750$

Promotion cost = 15% $\frac{15}{100} \times 3,75,000 = 56,250$

Transportation cost = 10%   = 37,500

Royalty cost = 10%   = 37,500

Total   Rs.3,75,000/-

(c) If the Marked Price Rs 25% above the Cost Price and if the MP is Rs.375/-, what is the cost of royalty in a single copy of the book?

What is the promotion cost on a single copy?

If cost of price = x; Marked Price is 1.25x

ie If MP is 1.25 x → CP is x

If MP is Rs.375/-, Cost Price = $\frac{x}{1.25x} \times 375 = 300$

Royalty = 10%

So the royalty is Rs.30/- per book on (CP)

Promotion cost = 10%, Rs.30/-

(d) If the royalty paid is Rs.3,00,000/- for an edition of 10,000 copies, what is the selling price if the publisher desires a profit of 5%?

If the CP is 100, for a profit of 5%; SP is 105 or 105% of CP. Let the selling price of this edition of 10,000 books be Rs.x/-

Then royalty 10% SP of a book in %

Total royalty = 3, 00, 000 : Total SP = x

$$r = \frac{105 \times 300000}{10} = 105 \times 3 \times 10^4$$

SP of one book is $\dfrac{105 \times 3 \times 10^4}{10000}$ = Rs.315/-

(e) If 10,000 copies of an edition are printed and the transportation cost on them is Rs. 3,00,000. What should be the selling price if the publisher wishes to earn a profit of 10%?

Selling Price - if profit desired is 10% ie 110% of CP

Transportation cost is 10%

Then 10: 110 = Rs.3,00,0000/- : x

Then $x = [(3 \times 10^5 \times 110)/10] = 33 \times 10^5$; No. of copies = 10,000

$$\therefore \text{ SP of each book} = \frac{33 \times 10^5}{10^4} = \text{Rs.330/-}$$

(f) By what percent is promotion cost less than printing cost?

Printing cost of the book = 20% of CP

Promotion cost of the book = 15% of CP

Difference = 5% of cp

$$\text{Percentage difference} = \left[\frac{\text{Actual difference}}{\text{Printing cost}} \times 100\right]$$

$$= \left[\frac{5\% \text{ of CP}}{20\% \text{ of CP}} \times 100\right]\% = 25\%$$

Thus promotion cost is 25% less than printing cost

41

Line Graphs

Line Graphs

Successive points representing the dependence of two variables are joined by straight lines in this. Analysing this variation, the questions are answered

Q.1 Exports of three organizations in different years are shown by line graphs. Answer the questions given.

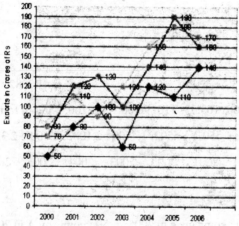

(Note: Tabulated Values Correct)
Exports in Crores of Rs.

Year → Org ↓	2000 (1)	2001 (2)	2002 (3)	2003 (4)	2004 (5)	2005 (6)	2006 (7)	Total (8)	Avg (9)
(1) A	50	80	110	70	120	110	140	680	97.143
(2) B	80	110	90	120	160	180	170	910	130
(3) C	70	120	130	100	140	190	160	910	130
(4) Total	200	310	330	290	420	480	470	2500	357.143
(5) Avg	66.66	103.33	110	96.66	140	160	156.66	833.33	119.048

1.A What is the relative proportion of average annual exports of A, B, C?

1.B What is the relative proportion of the seven average annual exports of all the 3 firms?

1.C What is the ratio of max. annual exports of any company to the minimum annual exports of any company in the period given?

1.D What is the ratio of max. annual exports of A to min annual exports of A?

1.E In how many years the annual exports of B are larger than the average annual exports of B?

1.F What is the difference of annual exports of all companies in 2006 to the annual exports of all companies 2000?

1.G In which year is the difference between exports of B and C the minimum?

Solutions from col. 9

1.A A   B   C   A : B   :   C
     97.143 : 130 : 130 = 1 : 1.3382 : 1.3382

[66.66: 103.33: 110: 96.66: 140: 160: 156.66] from col. 5

= [1: 1.5501; 1.6502; 1.4500 : 2.1002: 2.4002: 2.3501]

2000 2001 2002 2003 2004 2005 2006

1.B $\dfrac{190}{50} = \dfrac{19}{5} = 3.8$

(190 : company c in 2005;   50 : company A in 2000)

1.C $\dfrac{140}{50} = \dfrac{14}{5} = 2.8$; 140;   A in 2006;   50; A in 2000

1.D Average annual exports of B: 130 crores; In 2004, 2005, and 2006 the annual exports of B are larger (160, 180, 170) 3 years (from row 2)

1.E 156.66 - 66.66 = 90 crores (from row 5)

1.F Differences (in B, and C, + 10, - 10, - 40, 20, 20, - 10, 10 (from rows 2, 3)

(Say   B > C : + ve
       B < C : - ve)
       B → 90 crores (In 2002) ⎫
       C → 130 crores   ⎬ (-40)
       B < C   ⎭

Q.2 The numbers of students who joined a school and those who left the school at the beginning of the academic year is given for the period 2000 to 2006. Initial strength of the school in 1999 is 4000. Answer the given questions

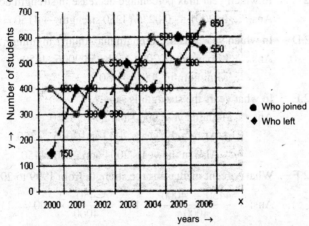

◆ Who left   ● Who joined, initial strength in 1999 is 4000

Table showing alterations in years; strength of college underlined (Row 2)

Sl. No.	2000		2001		2002		2003		2004		2005		2006	
	Joined	left	J	L	J	L	J	L	J	L	J	L	J	L
1	400	150	300	400	300	600	400	500	600	400	500	600	650	550
2	4400	4250	4550	4150	4650	4350	4750	4250	4850	4450	4950	4350	5000	4450
	Strength (Underlined in above line (2))	-	-	-	-	-	-	-	-	-	-	-	-	-
3	Increase (+) Decrease (-) in strength	+250 Compared to previous years' strength		-100		+200		-100		+200		-100		+100
4	percentage	0.0625		-0.0235		+0.0482		-0.0229		+0.0471		-0.0225		+0.0229

Explanation

Row: Strength of school in an year

Previous year's strength + no's joined - number's left

Note increase / decrease in strength relative to the previous year's strength are worked (row 3) out. Their percentage shown in (row 4)

2.A Show the strength of school in each year: Row 2
 Ans.: (underlined)

2.B In which year max. increase / decrease of strength took place?
 Ans.: In 2000 increase from 4000 to 4250 i.e., by 250. percentage increase is 0.0625

2.C In which year max percentage decrease in strength occurred
 Ans.: In 2001 : - 0.0235 (4250 strength → 4150 = -100)

2.D In which year was the strength maximum? minimum?
 Ans.: Maximum in 2004 & also in 2006: 4450
 Minimum in 2001 : 4150

2.E In what years the strength is same?
 Ans: a) In the year 2000 and 2003: 4250
 b) Also in the years 2002 and 2005 : 4350
 c) Also in the years 2004 and 2006: 4450

2.F What percent increase is the strength from 1999 to 2006?
 Ans: $\dfrac{4450 - 4000}{4000} \times 100 = \dfrac{450}{4000} \times 100 = 11.25\%$

2.G In which year there is the highest number of students joining?
 In 2006 : 650

2.H In which year maximum left the school in 2005 ? 600

2.I What is the ratio of least number of students who joined to the maxi number who left in any year (r) in the given period?
 Ans: Least number joined = 300 in 2001
 Maximum number left = 600 in 2005
 Ratio = $\dfrac{300}{600} = \dfrac{1}{2}$